Elmar Altvater · Achim Brunnengräber (Eds.)

After Cancún

VS RESEARCH

Energiepolitik und Klimaschutz

Herausgegeben von
PD Dr. Achim Brunnengräber, TU Dresden
PD Dr. Lutz Mez, FU Berlin

Elmar Altvater
Achim Brunnengräber (Eds.)

After Cancún

Climate Governance
or Climate Conflicts

VS RESEARCH

Bibliographic information published by the Deutsche Nationalbibliothek
The Deutsche Nationalbibliothek lists this publication in the Deutsche Nationalbibliografie;
detailed bibliographic data are available in the Internet at http://dnb.d-nb.de.

1st Edition 2011

Editorial Office: Frank Schindler | Verena Metzger

VS Verlag für Sozialwissenschaften is a brand of Springer Fachmedien.
Springer Fachmedien is part of Springer Science+Business Media.
www.vs-verlag.de

Cover design: KünkelLopka Medienentwicklung, Heidelberg
Printed on acid-free paper
Printed in Germany

ISBN 978-3-531-18291-9

Contents

Contributors

Elmar Altvater is a Professor emeritus for International Political Economy at the Free University Berlin and a member of the Scientific Council of ATTAC, Germany. E-Mail: altvater@zedat.fu-berlin.de

Achim Brunnengräber is a Visiting Professor at the Technische Universität Dresden, Chair for International Politics, and Associated Professor (Privatdozent) in Political Science at the Free University Berlin.
E-Mail: priklima@zedat.fu-berlin.de

Martin Bitter is a PhD student at the Free University Berlin. He is working on the emergence of a „European Carbon Economy" from a critical political economy approach. E-Mail: m-bitter@web.de

Bettina Knothe does research and consultancy in the field of sustainability and gender. She also works as a trainer in adult education with focus on gender trainings and teaches yoga. E-Mail: knothe@medeambiente.de

Lutz Mez is a Senior Associate Professor at the Department of Political and Social Sciences, Free University Berlin, and former managing director of the Environmental Policy Research Centre (FFU). E-Mail: lutz.mez@fu-berlin.de

Edward Nell is the Malcolm B. Smith Professor of Economics at the New School for Social Research in New York. E-Mail: nelle@newschool.edu

Peter Newell is currently Professor of Development Studies at the University of East Anglia. From September he will be Professor of International Relations at the University of Sussex. E-Mail: P.Newell@uea.ac.uk

Matthew Paterson is a Professor of Political Science at the University of Ottawa. E-Mail: Matthew.Paterson@uottawa.ca

Armon Rezai is an assistant professor in environmental economics at the Vienna University of Economics and Business (WU). E-Mail: Armon.Rezai@wu.ac.at

Miranda Schreurs is a Professor of Comparative Politics and the Director of the Environmental Policy Research Centre at the Free University Berlin. She is also Vice-Chair of the European Environment and Sustainable Development Advisory Councils (EEAC). E-Mail: miranda.schreurs@fu-berlin.de

Willi Semmler is Chair and Professor at the Department of Economics at the The New School, New York. E-Mail: SemmlerW@newschool.edu

Simon Wolf is a Professor of Economics at the New School for Research in New York. E-Mail: simonwolf@gmx.net

Preface

This volume is the final publication of a „Jointly Executed Research Project" (JERP) on questions of „Global Environmental Governance". It was one of 18 sub-projects within the Network of Excellence: „GARNET – Global Governance, Regionalisation and Regulation: The Role of the EU". The Network of Excellence was supported by the European Union within Research Framework Programme 6. It was coordinated by Richard Higgott, University of Warwick, UK. A number of working conferences and doctoral seminars were organised to coordinate the research activities of the scientists from several European countries participating in the JERP. The subjects ranged from European environmental governance and geopolitical aspects of environmental governance to money and finance in global environmental governance. In addition to a large number of individual publications the present anthology resulted from the workshops and scientific discussions. At the same time it represents the further development and updating of a German publication with the title „Ablasshandel gegen Klimawandel" (the selling of indulgences against climate change), published in 2008 by VSA Verlag in Hamburg.

No preface is complete without expressions of thanks: to the JERP partners, to the participants in the workshops and to the authors of this volume for their constructive collaboration. Bettina Knothe (Berlin) and Eleni Tsingou (Warwick) were untiring in the organisation and administration of the JERP. Alexander Wajnberg and Christin Linße painstakingly edited the texts. Irene Wilson provided excellent support in the translation and correction of a number of contributions. We also wish to express our thanks to Richard Higgott and his team at the University of Warwick, without whose commitment and cultivation of a cooperative atmosphere GARNET, the Network of Excellence, would have never come to existence.

Elmar Altvater and Achim Brunnengräber   Berlin, May 2011

With the Market Against Climate Catastrophe – Can That Succeed? – Introduction

Elmar Altvater / Achim Brunnengräber

The situation is paradoxical. The fossil-nuclear energy model causes catastrophes: at the end of the energy chain these are the emissions of greenhouse gases with their effects from the melting of the polar icecaps, the rising of the sea level, the flooding of coastal regions and the expansion of deserts, to the „unusual" weather events with heat waves and flooding which cost many people their lives (in Russia alone according to statistics of the reinsurance companies the number of deaths due to heat waves in 2010 was 55,000). These catastrophes influence the evolution of life, perhaps even hamper it, and they certainly cause considerable damage measured in monetary terms. The Stern-Report calculated this in 2006. The size of the losses to be expected is probably about 20 per cent of the global social product.

But already at the beginning of the fossil energy chain, in the exploration and production of non-conventional oil from the deep sea or from tar sand and oil shale, considerable damage to the environment takes place and there are repeated catastrophes such as the spectacular one caused by the explosion of the oil platform Deepwater Horizon in the spring of 2010, which caused contamination of the Gulf of Mexico. Less spectacular, although comparably damaging, are the contamination of lakes and rivers in the Niger delta and in the western Amazon, the extensive damage to the ecosystems of the Orinoko basin in Venezuela and the forests in Canadian Alberta. Non-conventional oil is focused on when conventional oil runs out. The peak of oil production („peak oil") has been reached and has perhaps already been passed. Wars are being conducted over access to the reserves and influence on price formation.

The risk of catastrophes is even greater in the nuclear cycle. Although this has a beginning when the uranium is extracted from the earth, so far it has no end, since there is no safe final disposal of nuclear waste. If the cycle is closed, then in the form of a catastrophe of immeasurable dimensions such as in Fukushima in March 2011. With nuclear energy the catastrophe is programmed as long as there is no place for the final disposal of waste for the next 100,000 years of human and geological history. Atomic catastrophe is only a matter of time, and that is what makes nuclear technology so dangerous. If, as statisticians emphasise, a worst case scenario should only take place every 10,000 years per reactor, i.e. only once in a period which covers the entire history of human civilisation from the beginning of culture in Mesopotamia to Fukushima, then with approximately 500 atomic reactors in the world today (but with an increasing trend) we can expect a worst case scenario every two decades. That is roughly the rhythm of Harrisburg 1979, Chernobyl 1986 and Fukushima 2011.

All these possible catastrophes, and those which have already become reality, point to the necessity of finding alternatives to both fossil and nuclear energy soon; that means the use of renewable energy sources. Probably, this can only be successful if at the same time the systems for the conversion of energy and the way in which these are used in production, consumption and transport are adapted, i.e. changed, because the „harvest" of renewable energy – related to the energy used to produce it (measured as EROEI = energy return on energy invested) – is as a rule less than in the case of fossil and nuclear energy. The energy concentrated in the fossil and nuclear resources of the Planet Earth allows these, furthermore, to be used independently of time and space, which is in stark contrast to the external energy source „sun", which is dependent on the time of day, the season and on the position on the „limited surface area" of the Planet Earth as well as on the regional situation. This is true in principle for all renewable energies: for wind, photovoltaics, water, tides, biomass, solar heat or geothermal energy.

The hardly controversial necessity of the transition to renewable energies is, however, weakened in the discourse. It is claimed that an increase in energy efficiency by the „factor 4" would suffice, and above all the transition to renewable energy would not require a fundamental change in societal relationships. The transformation of the energy regime seems possible without touching the relationships of power or the habits of the fossil-nuclear epoch. Theoretically, the revision or advancement of the proven theories of the academic mainstream is unnecessary and the accustomed thought patterns can be spun further and social practices continued.

The idea therefore is widespread that energy and climate policy can make use of market mechanisms, especially since the projects against climate change and for the introduction of alternative energy schemes have promising names: Green New Deal, Global Green Recovery, Green Climate Fund, Green Economy. Green is the colour of hope, and that this has its roots in the 20th century is expressed quite clearly in the reference to the New Deal of the Roosevelt administration in the USA of the 1930s. In the public debate the global environmental crisis and the impact of climate change on human beings has already morphed into a historical opportunity which should be grasped using „green" technology.

Growth has a future, despite the „limits to growth" which were pointed out by the Club of Rome in 1972. Jobs are created in the new industries of producing the equipment for alternative energies, a new basis of legitimation for development policy can be constructed with climate policy measures and even the crisis-ridden financial markets are helped with innovative futures for oil supplies or with certificates for emission rights which are traded on special exchanges. The vices of capitalism thus do not have to be left behind in the epoch of renewables, as energy and climate papers are magnificently suited to speculation. Market based financial instruments offer more than enough opportunities to make a profit with climate policy. The concept of reconciling climate policy with the market appears to be so charming and fascinating that it finds broad agreement not only in the scientific debate. Even some of the critics of „financially driven capitalism" can find advantages in the idea of opening up a new dynamic field of investment for idle capital. All at once the „limits to growth" can be politically overcome by reverting them into the new concepts of the „growth of the limits".

New fields of activity are opened up for economic and climate policy experiments undertaken by the old, but also by new, actors in international climate policy. Among these are not only new international organisations but also non-governmental organisations (NGOs) and transnational corporations. Their theoretical approaches are generally influenced by neoliberalism. World market conditions have been deeply changed by the flexibilisation of labour markets, by the opening of the markets for goods and services and last but not least by the liberalisation of financial markets, the privatisation of public goods and the far-reaching deregulation of policy. In this context, the application of the market mechanism as a tool to combat climate change finds many supporters among climate policy actors.

The pitfalls and limits of the market mechanism in climate policy are the subject of this book. It will be argued that the chances of a global green recovery of crisis-shaken capitalism can only be assessed and estimated against the background of the dominant neoliberal paradigm and by taking the existing functional logic of the system into account. The contributions in this volume show that climate policy is no longer determined only by the annual international climate negotiations (which in recent years failed again and again), but that climate change and climate protection have developed into a lucrative and integral part of a much broader financial market, trade and energy policy.

Although climate policy and climate science are still young, the discourses concerning them have already been realigned several times. Following the agreement on the United Nations Convention on Climate Change (UNCCC) in 1992 at the UN Conference on Environment and Development (UNCED) in Rio de Janeiro, climate change was treated as one of the most important global environmental problems (in addition to the threat to the diversity of species, the extension of deserts and the dramatic losses of forests). Five years later at the Conference in Kyoto the discussion of three market-based instruments with which the warming of the Earth's atmosphere was to be countered was in the foreground. These were, firstly, trade in CO_2 emission rights, secondly, „joint implementation" (JI), by which the reduction obligations in country A were to be lessened by investments to reduce emissions in country B, and thirdly the „clean development mechanism" (CDM), with which CO_2 reductions in developing countries were to be credited to companies in the industrial countries, if the latter invested in developing countries.

The idea of conducting climate policy with „market based instruments" met with widespread agreement from the beginning, but the doubts as to whether a radical reduction of CO_2 emissions can be achieved with market based instruments are now being voiced ever more loudly. Since 2008 the financial crisis has clearly demonstrated how little efficient financial markets are. Also, the implementation of the market based Kyoto instruments has not been free of conflict and friction. On the contrary, new problems have arisen, so that the question whether climate policy can be conducted successfully with „flexible instruments" continues to be the subject of controversial discussion and until now remains unanswered.

It is therefore all the more surprising that today hope is being spread that for the idle capital which has been lying fallow since the global financial and economic crisis investments to the tune of several thousand million US dollars for combating climate change would be possible; the struggle against climate change is thus being upgraded – at least discursively – to a comprehensive rescue package in order to overcome the financial turmoil. In view of the experiences with Deepwater Horizon and Fukushima at the beginning of the fossil and nuclear energy chain, however, scepticism is also growing as to whether market instruments and investments in green technology or in green investment funds will be able to achieve the obviously urgently needed change of direction in the energy and climate regimes.

The question from which we started, whether a Green New Deal appears realistic, therefore cannot be answered simply by looking at the Kyoto instruments. The functional logic and the structures of capitalism as a whole must be examined. Climate change threatens us all, although to differing degrees, and it represents a comprehensive crisis of societal relations of humankind to nature. A reduction in the emission of greenhouse gases must be achieved very quickly, more quickly than is being aimed for by the climate negotiations. The extent is also greater than the Kyoto protocol envisages. And according to all the forecasts for the consumption of fossil energy it appears almost impossible to achieve the target. 50% fewer CO_2 emissions by 2020 are necessary if the concentration of greenhouse gases in the atmosphere is to remain below the critical limit of 550 ppm. But how can that be achieved?

The Market: Your Friend and Helper?

There are only four possible paths. The first aims for an increase in energy efficiency in order to consume less fossil energy per unit of national product. The second path leads us into landscapes of carbon sinks, primarily in re-afforested woods and monocultures in the global South. The market based instruments are intended (on the first path) to sink emissions via an increase in efficiency in the use of energy, and on the second path with the help of ET, CDM and JI to provide for the reduction of emitted CO_2.

On the third path the emitted CO_2 is separated, captured and stored in the Earth's crust (carbon capture and storage, CCS). It is not sure whether this method is technically viable. Only the fourth path leads away from the fossil energy regime to a world of renewable, solar energy sources, i.e. away from the closed fossil and nuclear energy systems into the open, solar energy system. In the closed system at the beginning of the energy chain the energy sources are extracted from the Earth's crust and at the end of the energy chain the emissions (greenhouse gases and atomic waste) are „disposed of" in the atmosphere or the pedosphere and lithosphere of the Earth. After the transition to the solar energy regime the remaining fossil reserves can stay in the earth: „leave the oil in the soil". Which path will be taken is a question of hegemonic struggles and of political decisions. These can foresee market incentives, or they can be based on permits and prohibitions, on active state investments and the control of these, but also on enlightening political education. In the Kyoto agreement the commitment is above all to the incentive system of the market.

This is paradoxical, firstly because a market for CO_2 does not exist. CO_2 has no use-value with which needs could be satisfied. On the contrary, it is harmful; it therefore cannot be transformed into a tradable commodity. CO_2 is a bad, not a good, which one would wish to get rid of as quickly as possible – if only that were so easy. Since there is no market for CO_2, it is impossible to regulate its exchange effectively by using market mechanisms. The obvious thing to do, therefore, would be to prevent CO_2 emissions by means of legal requirements and prohibitions, with threshold values and technical prescriptions.

However, it is possible to exchange pollution rights on a market which must be created by the state. Admittedly, the „making" of a market requires considerable preconditions. Although the atmosphere, into which the greenhouse gases are released, is not privatised, and CO_2 does not become a private asset, rights to the pollution of the atmosphere („allowances") are politically constructed. These are then awarded to CO_2 emitters according to a national allocation plan – gratuitously, as in the EU to date, or for a price determined by auction. This will possibly be the case in the EU as of 2012. The scarcity of the economic good „pollution rights" is therefore determined artificially, i.e. politically, namely by the upper limits to emissions („caps"). „Green climate capitalism" is thus only so charming because it is politicised through and through. Something not really tradable must be transformed into a tradable good. This is a political trick by which, however, things are given their real nature, namely to be objects tradable by private persons.

The producers of CO_2 now have an individual economic right to pollute the atmosphere. They receive a politically certified good which they can trade like sides of bacon, barrels of oil, Christmas decorations or stocks, option certificates, futures and other financial instruments which are „originated" by securitisation. But certificate markets do not function in the same way as weekly markets at which people not only buy and sell but also like to stop for a chat. They have a global reach, they are under the influence of power, they are subject to hard locational competition and are drawn into the intrigues on the financial markets and into their tendencies to crisis; they land in the financial speculators' arsenal of „financial weapons of mass destruction" (Warren Buffet). Price movements on an artificial market such as that for emission certificates are erratic and extremely volatile, as *Nell, Semmler and Rezai* show in this volume. The certificates have no connection to costs of labour and capital because the prices are generated on a market which has no history, no morals values and no assignable costs. A market without history is like a rootless reed in the wind, however, and therefore the high volatility is not surprising.

The market based instruments of climate protection do look elegant, however. They fit into the world view of a global (neo)liberal order, in which market ranks above planning, economics above politics and the private sector above public goods and the state. Many environmentalists, globalisation critics, representatives of green and left-wing parties and the majority of environmental economists have fallen for the charm exuded by market economy solutions even when they fail. They are fascinated by the promised artifice of an idea: quantities are determined (caps) politically and then the free market mechanism generates price signals and profit incentives in such a way that the pursuit of individual interests leads to an optimal result for everybody, for the entirety of the more than six billion citizens of the Earth. The result of applying this intelligent mechanism is a reduction in greenhouse gas emissions by the percentage which is necessary from a climate policy perspective – without legal requirements, prohibitions and state bureaucracy, with all the freedom of the market.

Nevertheless, a further consideration raises fundamental doubts as to the suitability of market economy instruments for combating climate change. Markets are economic places where the supply of goods meets with monetary demand. A market economy is therefore always (*per definitionem*) a monetary economy, whether the money in circulation is termed Dollar, Euro, Peso or Kauri. If money becomes a commodity, then financial markets are formed. Money and financial assets are traded at a market price. This is interest rate. Interest must be paid, however, by those demanding money, by investors, to those granting credit, e.g. banks. The satisfaction of interest demands is only possible, however, if a real surplus, a surplus value is produced, i.e. if the economy grows. This only takes place if additional materials and energy are consumed and, accordingly, if additional greenhouse gases are emitted. Instead of contributing to the reduction of greenhouse gas emissions, market instruments promote them.

The market economy system, and its dynamism and efficiency which are so praised by its neoliberal supporters, is a precondition for a market based climate policy and at the same time is supporting it. Much can be expected from market based instruments – but not a radical reduction of greenhouse gas emissions. The reduction eventually takes place during and as a result of the crises of the market. Thus, if the market mechanism cannot be trusted, environmental taxes (e.g. a *carbon tax*) and legal regulations represent a better and more adequate means to reduce greenhouse gas emissions. In addition – on the fourth path – a socio-ecological reconstruction in the direction of a solar society must become the most important environmental policy objective.

On the Contributions in this Volume

Peter Newell and **Matthew Paterson** demonstrate that the political answers to climate change are developed in the context of a global, neoliberal capitalism, and ask under what conditions decarbonisation would be possible. They identify strategies and debates which would already lead to tensions and contradictions with the present fossil capitalist system. In spite of the obvious problems and difficulties of these approaches certain social powers challenge capitalism and its hegemony in the course of a socio-techno-political transformation. With reference to Gramsci, they theoretically assume that there now exists at least a historic bloc in construction which they can imagine sustaining climate change policy. Their argument is discussed taking the examples of the emissions trade as it is practised in the European Union and of the Clean Development Mechanism. Newell and

Paterson refer to, but do not discuss in detail, the increasing importance of the „financialisation of nature", or what Marx called „fictitious capital".

Simon Wolf argues that climate politics has turned into an issue of finance and investment, as scaling up the financial support for mitigation in developing countries is said to be crucial for successfully meeting the climate challenge. Informed by an economic understanding of climate change that was primarily established by the Stern Review, the climate finance discourse compares abatement opportunities across sectors and world regions along their cost-effectiveness, and concludes that a large share of global emission reductions must happen in developing countries. One consequence of this rationality is the urgency that is given to implementing a financial mechanism for reducing emissions from deforestation and degradation (REDD). As the larger share of developing countries in global emission reductions requires levels of climate finance that by far exceed the resources currently available from the funding mechanisms under the UNFCCC and beyond, a central role is ascribed to private investments. The role of governments, accordingly, is to incentivise private finance flows by creating conducive investment environments and overcoming investment barriers through regulation and public finance mechanisms.

An alternative framing of the climate finance issue that started from political objectives rather than economic rationalities and constraints, Simon Wolf argues, would reveal that traditional forms of regulation such as standard setting or taxes could be used for redirecting existing financial flows rather than hectically searching for new sources of funding. This would, at the same time, not only address some of the root causes of climate change, but provide governments with a new stream of income that could be used for policies and measures that are not deemed attractive by private investors, but are desirable from a societal point of view.

The „ecological modernisation" approach, writes **Martin Bitter,** is based on the assumption that capital accumulation could be politically regulated in a way that cuts pollution to an „optimal level". Its basic strategy is the valorisation and monetisation of nature, commonly known as the „internalisation of external effects". In his article, however, it is argued that nature and economy have to be grasped from an immanent perspective in order to understand the contradictory character of this internalisation. Applied to the case of the European Union Emissions Trading Scheme (EU ETS), it is conceded that carbon is being priced while becoming an important factor of production. However, this process implies contradictory effects because a) a political construction of (temporary) property rights is necessary which tends to perpetuate social power relations from the Fordist Epoch; b) the monetary expression of carbon is inherently arbitrary and prone to crisis; c) the individualised natural commodity conceals the underlying social relations that are the driving forces of climate change; d) the transformation of different socio-natural relations into the exchange value of CO_2 equivalents abstracts from important structural distinctions in the society-nature metabolism. The article concludes with the enhancement of different time structures of (financial) markets, policy making and nature, stressing their „structural causality" (Althusser) for future investigations of carbon markets.

In the contribution by **Edward Nell, Willi Semmler** and **Armon Rezai**, the authors discuss economic theories with which emissions trade is justified and compare them with theories preferring to combat the global rise in temperatures by means of taxation. In unison with the IPCC reports they show that the regulation via taxes is preferable to market based instruments. The experiences of the European emissions trade system are presented as an argument in order to show empirically that market based instruments are disappointing. None of the objectives aimed for could be achieved. The volatility of the prices of certificates is extremely high, so that this method even tends to have a counterproductive effect with regard to environmental policy.

Achim Brunnengräber discusses the climate policy of the EU. He examines explicitly the indissoluble connection between energy policy and climate policy, focusing on the energy chain from the inputs to the outputs, the CO_2 emissions. In order to counter the political and economic risks which arise from the dependence of the EU on energy imports, the EU wishes to start a „new industrial revolution" to speed up low-carbon growth, to increase its own energy production dramatically and to maximise competitiveness. The expansion of renewable energy forms and the flexible instruments of the Kyoto protocol, especially emissions trade, are part of this strategical approach. At the same time, however, the EU was unable to effect the strengthening of renewable energies in the international climate negotiations nor could it establish a strategy oriented towards the internal market which could have prevented the fact that the Kyoto objective of the EU (-8% by 2012 compared to 1990) will only be achievable with the aid of „escape routes" (sinks, CDM). At the same time EU energy security policy aims to secure the supply of oil and gas, to win new regions for this (import diversification) and to make investments in new and better pipelines and storage. This shows the strong obsession of the EU with low energy prices as a precondition for economic growth and for the realisation of the Lisbon strategy of 2005. Cheap and reliable energy provision is necessary for the realisation of the „Global Europe"- concept, of 2007, namely to morph into the most competitive and dynamic economic area in the world.

Elmar Altvater argues that the energy chain is doubly determined, from the extraction of the reserves of fossil fuels to the emissions caused by their combustion in order to gain useful working energy. We are dealing with material and energy transformations which are the subject of thermodynamic economics and of natural science and at the same time with money and capital. Firstly, the monetary value of the reserves is estimated. This determines the market value of oil firms or the creditworthiness of governments. Secondly, the oil is traded as *wet oil* at prices formed on markets and as „paper oil", e.g. as oil futures, on exchanges. Thirdly, at the end of the energy chain CO_2 is emitted into the Earth's atmosphere – with the known consequences. At the same time tradable pollution rights, which can be traded for money on special markets, are securitised and thus „originated". In fossil capitalism the fossil energy chain thus doubles itself into a carbon cycle and a value cycle.

Is the considerable reduction of emissions of greenhouse gases at the end of the fossil energy chain in the coming decades possible – 50% until 2050 or 80% until the end of the century by intervening with market based instruments into the value cycle and not directly into the carbon cycle? With the instrument of emissions trade the carbon cycle can possibly be regulated among dumping grounds (CO_2 in the atmosphere), sinks (via the CDM) and storage (CCS), i.e. it can make progress along the first three paths, but it cannot regulate the interface between fossil energy sources and their release after they have been produced; the fourth path is blocked. Emissions trade is unsuitable for corking the bottle from which the spirit of CO_2 is escaping.

It is nevertheless necessary and meaningful to take the experiences with the emissions trading system into account. **Miranda Schreurs** describes the experiences with the emissions trading system in the USA. She shows that its efficiency depends on which emissions are involved, how the system is constructed and how the institutions of the trading system function. It is also important that civil society organisations examine and observe the system, not least in order to clarify issues of environmental justice. She demonstrates the theoretical assumptions on which emissions trade is based, discusses its introduction within the framework of the United States Environmental Protection Agency's 1974 Emission Trading Program and the Clean Air Act Program. The mechanism of emissions trade was then extended to manage local air pollution, water pollution etc. Emissions trade thus has a long tradition in the USA and this is one of the reasons why the principle was also anchored in the Kyoto protocol as a flexible mechanism – even though neither the Bush nor the Obama government have signed the protocol. It is also shown in the article that systems of emissions trade have been introduced in transnational concerns.

The article by **Bettina Knothe** discusses the thesis that the success of internal implementation and external cooperation within European environmental policy is strongly influenced by the *quality* of the negotiations on the use of common pool resources. Using the example of water resources management under the European Water Framework Directive she argues that efforts towards the transparent governance of natural resources still remain in dichotomised patterns of societal apperceptions and political actions. From a socio-ecologic research perspective these patterns are built upon hegemonic relations between society and nature which hinder the implementation of an equitable, environmentally adjusted and socially sound supply and care economy.

Linking critical feminist theory and gender research with strong sustainability approaches, the article develops the idea that a precondition for new forms of political alliance which support the breakup of categorical and stereotypical dualities and functional attributions in environmental regulation may consist of two factors: the constructive acknowledgement of the intense relation of the existence of human life to its ecological environment as an existential resource on the one hand, and the intersecting factors which are responsible for identity-establishing and society formation on the other. Acknowledging the ambivalence for daily-life practices being societal and economically rooted in both global and local constraints, the argumentation on the role of civil society in natural resources supply chains focuses on the concept of „shared meaningful intersubjective spaces" as a common interpersonal sphere of individual meaning and social seriousness in erecting cultural systems, surroundings and rationalities. This perspective is chosen to connect the productive and reproductive spheres of supply chains by putting „care", „responsibility" and „precaution" at the centre of socio-political decision-making within natural resources management.

Lutz Mez and **Achim Brunnengräber** examine the „visions" of a sustainable energy supply and discuss the prospects. If greenhouse gas emissions worldwide are to be more than halved by 2050 compared to the situation in 1990, then, firstly, energy efficiency (the rational transformation and use of energy and electricity) must be increased, and secondly, fossil (coal, oil and gas) and nuclear energy systems must be replaced by an environmentally friendly, renewable energy system. The potentials of renewable energy are large. They can contribute to the preservation of resources, to the struggle against poverty in the developing countries and to the substitution of fossil energy imports into the industrial countries or to the avoidance of conflicts with regard to the ever scarcer fossil energy resources. Renewable energies, in addition, open up opportunities for decentralised energy use and create new jobs and occupations. This contribution describes and analyses the different support instruments which are being applied to the expansion of renewable energies in Europe and the world. And it argues with respect to current studies that the necessary phase-out of nuclear energy, which following Fukushima cannot be halted, would be rapidly possible with the aid of energy efficiency and renewables; much more rapidly than was long assumed to be the case.

Climate Capitalism[1]

Peter Newell / Matthew Paterson

Whether we like it or not, for the foreseeable future at least, responses to climate change will be developed in a context of global capitalism. What does this imply in political and strategic terms and how can we make sense of this theoretically? In this chapter we try to nuance eco-Marxist claims which might result from this more or less banal, if perhaps depressing, observation about capitalism's hold over climate politics. There is now a substantial and rapidly growing such literature on climate politics, which has increasingly focused on the commodification of climate change, through the establishment of markets in emissions allowances or credits, and which is, reasonably enough, highly critical of such commodification processes (Prudham 2009; Castree 2003; Lohmann 2005). There are also a number of analyses rehearsing well-known ‚second contradiction of capitalism' (O'Connor 1994; Kovell 2002; Benton 2000 Sandler 1994) or ‚treadmill of production' arguments (Roberts et al 2003; Gould et al 2008) in relation to climate change – suggesting that capitalism's growth-addiction and fossil fuel dependence means that it cannot possibly decarbonise. We seek here to nuance the depressing conclusions that can be drawn from such analyses and ask under what conditions might we be able to imagine capitalism decarbonising?

Our starting point for such an analysis is the observation that judged by the number of initiatives and levels of finance now being committed, many capitalists and state elites, for a range of different reasons, now have a political and financial stake in the project of decarbonisation. To be sure, there is a whole swathe of greenwash to wade through, but it is nevertheless the case that a tangible constituency of ‚climate capitalists' with a material interest in decarbonisation exists. They may currently be relatively marginal to business-as-usual models of capitalism, but it is worth posing the question about whether and under what conditions a carbon economy might develop into a project of climate capitalism in which growth is achieved through low carbon development.

---

[1] This chapter is an elaboration, with significantly more theoretical engagement, of the main arguments of our book *Climate capitalism: global warming and the transformation of the global economy* (Newell and Paterson 2010). That book was written for a broad popular audience and thus contains little of the explicit theorisations we try to develop here.

Projects to decarbonise capitalism don't of course emerge from a void. They rather start from a particular moment in time with its attendant structural features. Climate politics reflects both the structural features of capitalism in general and the legacy of neo-liberalism in particular because it is as a result of these structures, institutions and practices that responses to climate change have started to emerge. It is also the case, however, that the imperative of addressing climate change and moving towards low carbon development paths require new forms of capitalism. What this means politically is that strategies and actions to tackle climate change have to gain traction with social forces and actors that wield significant power within contemporary forms of capitalism. That this will be a contradictory, messy, and problematic process is obvious; but our working contention here is that we ought to take seriously that it might not be impossible.

In this chapter we look first at some of the principal characteristics and features of capitalism in general and of neo-liberalism in particular that are pertinent to, and structure the opportunities for, action on climate change. Second, we look at the ways in which some ways of tackling climate change have won out, been ‚naturalised‘ and come to be seen as more effective and legitimate ways of responding to the issue than others. We do this through two short case studies of emissions trading and carbon offset markets. Third and finally we explore ways in which we might understand these dynamics theoretically.

1   Capitalism and Climate Change

Perhaps more than any other environmental issue, climate change appears to place in stark relief the limits of capitalism's ability to deal with the environmental problems that it generates. Whereas new markets and products could relatively quickly and easily be created for alternatives to ozone-depleting chemicals such as CFCs, and technological fixes were available to address acid rain, the centrality of energy use based on widespread use of fossil fuels means action on climate change potentially disrupts capitalism's ability, indeed necessity, to grow. Hence while strategies of ecological modernisation demonstrate the ability of capitalism to deliver growth without increased emissions, through innovative use of technology and realising efficiency gains, tackling climate change means producing and consuming less of the very sources of energy that have fuelled the industrial revolution, colonialism and post-war economic growth. It is for this reason that many have suggested that addressing climate change and continuing with capitalism are mutually exclusive: you cannot have both (Kovell 2002). Much of this claim is based on the centrality of fossil fuels, and especially the properties of oil, to capitalism (Altvater 2006: 39-41). Whether climate change reflects a ‚crisis of the capitalist mode of production‘ as Brunnengräber (2006: 219) claims, is a moot point.

At the very least, though, thinking about climate change in a capitalist context entails a project of broad socio-techno-political transformation. Capitalism as usual cannot deliver the scale of change required to tackle climate change and hence needs to transform itself once again. But of course capitalism has engendered such transformations before, albeit perhaps on a different scale and not always by intentional design. Popular reference points for the nature of the transformation required might include the industrial revolution (and a need for a second industrial revolution), a new green deal (modelled on the Marshall Plan or the Bretton Woods institutions in terms of scale of required finance as well as the need for new global governance institutions) or widespread socio-technical transitions such as the development and roll-out of the railways during the 19th century (Newell and Paterson 2010). The point to note here is that fossil fuel capitalism is just one form, albeit an entrenched and enduring one, of capitalism. While certain ‚base technologies‘ (Storper and Walker 1989) may characterise eras of capitalism, as Buck notes, it is important not to ‚confuse particular manifestations of capitalism- that is, particular historical social formations- with capitalism itself, thus under-estimating the flexibility of the beast‘ (2006: 60). Even a post-oil economy, he argues, would be a capitalist one as long as there is an industrial reserve army without ownership or control of the means of production and as long as the production of commodities by commodities prevails. Hence, even peak oil can be re-worked as an opportunity for growth where fossil fuels can be replaced by a ‚solar revolution‘ (Altvater 2006: 53). As Marx and Engels famously stated, the bourgeoisie ‚cannot exist without constantly revolutionizing the means of production‘ (1978: 28). Technological dynamism is at the heart of capitalism, and as a consequence, its technological trajectories are not necessarily set in stone. Capital, as value in motion, does not care about what it makes, the machinery used or the motive source. It cares only about its own self-expansion and valorisation (Buck 2006: 63). These are the incessant waves of creative destruction that need to be harnessed towards the goal of a low carbon economy.

Responses to climate change are most immediately structured by the key features of neo-liberalism, and we suggest here that four, in particular, are crucial. There are few surprises here. First is the dominance of pro-market ideology, the rhetoric that markets are the most ‚natural‘ sort of social organisation (Gamble 1988; Frank 2000). In the context of climate change and environmental debates more generally, this has led to an ideological undermining of direct regulatory (so-called ‚command and control‘) approaches in favour either of fiscal instruments like carbon taxes or, even better for market purists, emissions trading systems where private property rights in environmental services are established.

That this has favoured the dominance of market-based approaches, notably emissions trading and carbon offset markets that we elaborate in more detail below, is fairly obvious. As we have argued elsewhere:

> Had climate change become seen as an acute problem in 1950 rather than from the late 1980s onwards, anyone proposing emissions trading as the principal means to deal with it would have been subject to ridicule ... The rise of the carbon economy, as one manifestation of a broader trend towards the ‚marketisation' of environmental governance, evolves alongside and is a product of the entrenchment of neo-liberal politics throughout the 1990s (Newell and Paterson 2009: 77).

But market ideology has not been sufficient. A second key feature of neo-liberalism is the structural importance of finance within neo-liberalism. Many talk of a ‚finance-led regime of accumulation' as the distinctive economic dynamic of neo-liberalism (Boyer 2000; Aglietta 1999), and the dominance of finance in the contemporary economy is widely recognised (Helleiner 1994; Leyshon and Thrift 1997). But there is a broad consensus across a range of perspectives that what was distinctive about the period from around 1980 was the re-birth of a global finance that gained both extreme mobility and thus capacity to discipline states, but also enhanced power over other fractions of capital, notably manufacturing (Cox 1994). This heightened power of finance is important in understanding the popularity of carbon markets, since financiers are the primary beneficiary of this particular policy design.[2] But it also helps understand other features of climate governance – notably the emergence of investor networks like the Carbon Disclosure Project (CDP) that attempt to improve information flows about firms' exposure to climate change impacts and their carbon-intensity and thus vulnerability to emissions reductions policies.

Third are the soaring inequalities provoked by neo-liberal forms of economic management. Within countries, such inequalities have sharpened almost everywhere since the early 1980s, through combinations of increased unemployment, welfare retrenchment, and more unequally distributed tax burdens. Globally, they have sharpened because of monetary policies fixated on the use of interest rates, the shift in value-added from manufacturing to financial services, substitution of key resources of developing country exports, and the like. Such inequalities have both been the basis for the highly conflictual character of global

---

[2] It is worth noting, however, that financiers play an extremely small role in the direct design of policy. Apart from in the recent pressure on the Clean Development Mechanism process to become ‚less bureaucratic', mostly channelled through associations such as the IETA (International Emissions Trading Association), they are striking in their absence from lobbying or consultation in policy processes. Nevertheless, their structural role is crucial in understanding the success of carbon markets as policy instruments.

climate negotiations since their inception in 1991 (Roberts and Park 2007), but also generated opportunities for the exploitation and commodification of the basis of that conflict. The emergence of the Clean Development Mechanism (CDM) in the Kyoto Protocol can be interpreted from this point of view precisely as the institutional means by which to forge a compromise between the two broad groups of states with differentiated responsibility for addressing the issue but uneven opportunities to do so. It enabled a de-politicisation of the resolution of the North-South conflict by redistributing the ‚burden' of emissions reductions from North to South through a market mechanism which could be presented as apolitical, and even beneficial to certain Southern actors. As we will see below, adding ‚sustainable development' as a criteria which emissions reductions projects paid for by the global North had to meet, was supposed to be a way of ensuring development benefits flowed from making cheaper emissions reductions available to richer countries.

Finally, neo-liberal economies are often understood (if rhetorically, even ideologically) as about fast-moving networks replacing old bureaucratic (state and corporate) hierarchies. In the climate change context, it is possible to find all sorts of examples of this sort of organisational arrangement. It can variously be seen: in the shift to ‚governance', voluntary agreements and partnerships between states and corporations; in the swathe of private sector governance initiatives, such as the CDP already mentioned, the various certification standards that attempt to govern the voluntary carbon market, or the adoption of climate change measures by corporations under the rubric of Corporate Social Responsibility (Bulkeley and Newell 2010). It can even be observed in the character of social movement mobilising by groups such as Plane Stupid or the Climate Justice movement to promote radical climate action and to resist carbon markets.

So the pursuit of ‚climate capitalism' is strongly shaped by the neo-liberal context in which it emerged, with all its problems and contradictions. A number of features of this carbon economy result. First, the role of the state in relation to climate change has been configured around facilitating and enabling a series of possibilities for commodification and accumulation for various parts of capital, in particular for financiers and closely associated sectors (auditors, lawyers, especially). International institutions such as the World Bank have also been closely involved in promoting this sort of carbon economy – with the Prototype Carbon Fund and a number of other Bank facilities being established to promote investment in the CDM and, more recently, to deliver carbon finance in the form of the Climate Investment Funds (World Bank 2008).

Second, finance has been given a long leash in inventing a ‚carbon economy' out of the specific policy instruments. We, therefore, do not only have primary trading in the allowances and credits created by particular schemes like the

EU ETS or the CDM, but we also have increasingly elaborate derivative instruments (although as yet nowhere near as complex as asset-backed securities, for example), with futures and options markets enabling regulated firms to hedge against price volatility as other firms do in other financial markets (airlines hedging oil price volatility, for example, or farmers and agribusiness hedging grain price volatility). We also have an elaborate set of secondary commodification processes – auditors selling services to verify claims made in CDM project documents, insurers such as Munich Re providing services to insure against a CDM project not being approved, and so on. Finally, there are other financial instruments emerging, only tangentially connected to climate but perhaps stimulated by the enhanced sensitivity to weather/climate questions, such as weather derivatives enabling various institutions (most famously, pubs and restaurants) to hedge against adverse weather (Pryke 2007; Pollard et al 2008).

Third, this sort of response to climate change entails various exploitative processes endemic to capitalism and intensified under neo-liberalism. Bumpus and Liverman (2008) have borrowed from Harvey (2005) in coining the term ‚accumulation by decarbonisation‘ to capture this process. While decarbonisation is more commonly referred to as the process of taking the carbon out of the global economy, they use it to refer to processes of taking other people's carbon (rights) away from them, a slightly more precise term than the ‚carbon colonialism‘ prevalent in critiques of carbon markets (Bachram 2004, Lohmann 2006). Two specific processes are involved. First, in international negotiations, unequal rights to emit are being written into legal documents like the Kyoto Protocol, and these rights then turned into commodities. So rich countries are gaining assets in the process, assets that are not available to poorer countries. Second, in the CDM in particular, but also in the voluntary carbon markets, this entails a process whereby the rich are able to invest in developing countries to offset their emissions, in practice therefore taking away future possible emissions rights that people in those countries ought to have.

We should emphasise at this stage that identifying and elaborating upon these trends is not intended to imply that there is one overarching homogeneous carbon economy (Castree 2006) or organising force (Boykoff and Randalls 2009). Indeed, considerable effort is invested in trying to link different elements of the carbon economy, created as they are for different purposes and embedded as they are in distinct governance arrangements (Newell and Paterson 2010). This is the case with the EU's linking directive which ties the ETS to the CDM, for example. It is also the case that the emergent forms of climate capitalism we describe here reflect tensions and struggle between different fractions of capital and social forces to ensure that responses to climate change maximise their own opportunities for further accumulation.[3] Below we describe in more detail the features of two specific sites of accumulation in the carbon economy.

2   New sites of accumulation: Emissions trading and offsets

Given the neo-liberal context within which it arose, central to the response to climate change has been the process of commodification (Castree 2003; Smith 2006; Prudham 2009). It is often argued that the new generation of ecological commodities, including carbon, is different from earlier exploitation of the eco-logical conditions of production, being based not on the use values of resources themselves, but on ‚a major strategy for ecological commodification, marketiza-tion and financialization which radically intensifies and deepens the penetration of nature by capital‘ (Smith 2006: 17, cf. the contribution by Wolf in this vol-ume). What is interesting with carbon is that its worth, expressed in tonnes, rests precisely on the fact it cannot be productively consumed – it is avoided emis-sions that are paid for and traded on markets.

Here we elaborate two such efforts at carbon commodification: the Europe-an Union Emissions Trading Scheme (EU ETS) and the Clean Development Mechanism (CDM). These illustrate the logic and process well of this sort of response, but also are currently the most important such markets, accounting for around 80% of the volume of trades in contemporary carbon markets (Capoor and Ambrosi 2009). Conceptually, they highlight the two distinctive sorts of markets and thus commodification processes that have emerged – emissions trading, or cap and trade markets on the one hand, and offset, or baseline and credit markets, on the other.

Emissions Trading

The EU ETS is a paradigmatic example of an emissions trading system. Ideolog-ically, it reflects a purist neoclassical economic logic. In this logic, climate change exists fundamentally as a problem of open access resources (the classi-cally misnamed commons problems of Garrett Hardin). Because no-one ‚owns‘ the resource, no-one has an incentive to use it wisely. The response is thus to

---

[3] For example, Smith (2006: 32-33) describes the way in which ‚The US rejection of the Kyoto Protocol represents an internal ruling class jostle between more environmentally ‚friendly‘ energy capitalists ...and more aggressive cowboy capitalists who, while quite happy to invest in the envi-ronmental market, see their immediate profits in terms of direct energy production for an expanding market‘. This is similar to Dalby and Paterson's (2009) description of a split between a ‚carbonifer-ous capitalist‘ bloc and an ‚ecological modernisation‘ bloc.

privatise the resource, to create a series of private property rights in the resources' use, to create the incentives not to overuse it. The second part of the response is then to create a market in these property rights, so that the appropriate signals about the relative costs of different ways of living within this resource scarcity (in this case, ways to reduce GHG emissions) are set by the market itself rather than by government regulators (who set up the market and establish the basic constraint of the number of available permits).

The EU ETS does so by regulating individual installations that emit CO_2. It regulates around 11000 such installations in the EU, representing around 45% of the EU's total emissions. It started in a pilot phase (phase I), which ran from 2005-07. It is currently in its 2nd phase which runs concurrently with the Kyoto Protocol's commitment period, 2008-12, and will have a 3rd phase from 2013-20. In each phase, there is a process whereby an overall emissions cap is set at the EU level. Up to now, this is then allocated amongst member states who allocate allowances to specific installations. As of 2013, the Commission will directly allocate to the installations. Each installation is then required to hold the number of allowances equivalent to their emissions, at the end of the specified period. If they do not have enough to cover them, they have to buy in extra allowances. If a company can, therefore, reduce emissions below its set amount, it has allowances to sell within the system. The resulting trading creates a ‚carbon price' that acts as a signal to companies regarding what level of abatement it is rational for them to pursue.

In principle, this means that there is a tight cap on emissions, combined with a mechanism that means that emissions are reduced where it is cheapest to do so. All firms face incentives to limit emissions, but those incentives will be compared to the costs of investing to reduce emissions, resulting in different strategies amongst firms. But of course in practice there are a number of problems. First is the problem of information asymmetry. At present, companies are for the most part given the allowances free, and thus they negotiate with regulators how many constitutes a fair division of the overall burden for them. But they have much better information about the costs to them of reducing emission than do regulators, so there is plenty of potential for gaming, with firms ending up with more allowances than they need to start with. Electricity generators in particular have ended up getting significant windfall profits from this process, being given more allowances than they needed, passing on any costs to consumers while selling excess allowances into the system (Sandbag 2010; Gilbertson and Reyes 2009). Even without this windfall profits problem, this information asymmetry means that the allocations have generally been more generous than they should have been, which caused, among other things, a (temporary) price collapse in the EU allowance price in late 2006 as it became clear (with new, more precise information about the baseline emissions in the installations) that the allocations were ,long'.

Second is that in 2004, as the final rules for the system were being established, the EU decided to allow firms to buy in credits from the CDM (on which see below) to count against their obligations under the EU ETS. This ,linking directive' has arguably significantly watered down the emissions reductions potential of the ETS, as it means in effect that the overall cap is not a strict cap any more; emissions can be reduced by less than the cap requires (or even perhaps grow) and instead be offset through investments in the CDM. There are limits on how much of their obligations firms can meet through CDM credits, but nevertheless this constitutes a weakening of the strict claims that the ETS is cutting emissions in the EU.

Third is that the financial markets are distinctly under-regulated. In part this is because of their novelty; it was difficult to conceptualise the problems that may arise in the operations of this new sort of market. But it is mostly because of the neo-liberal heritage, which favours light regulation of such markets. This has, however, produced a number of ongoing problems.[4] The accountancy governance system has so far failed to decide on how the allowances, or any other carbon instrument, should be treated in company accounts – crucially, whether and when they should appear as assets or liabilities in financial accounts. This clearly leaves significant room for different strategies by firms to report for tax purposes, depending on their interests. Moreover, regulators have not really decided whether or not carbon markets are in fact financial markets, and thus whether market regulation should be dealt with by regular financial regulators or should be kept as part of the environmental regulatory arrangements.[5]

Opinions on whether the EU ETS has in fact helped reduce emissions vary considerably. Critics of carbon markets maintain that its impact is minimal at best, and at worst simply mask ongoing emissions increases, especially because of the availability of offsets (Sandbag 2010: 14). The World Bank, by contrast, argues that the EU ETS, even given problems to do with over-allocation and the availability of offsets, in fact has contributed to a modest emissions reduction amongst the regulated firms, and anticipates better performance with more stringent targets, direct allocation by the Commission, and significantly increased

---

[4] One problem which we leave to a footnote here is the tax scams currently being investigated by Europol (Philips 2009). These seem to us in fact just VAT tax scams and there is little distinctive about the ETS that makes it uniquely a problem of the carbon market *per se*.

[5] This problem is being repeated in the US, regarding whether the cap and trade markets that will result from legislation coming out of Washington (if it succeeds) should be regulated by the EPA or the Commodity Futures Trading Commission.

auctioning rather than free allocation of the allowances (Capoor and Ambrosi 2009).

Whether or not this is (yet) the case, a political-economic analysis of the EU ETS makes it clear that as a policy tool, it has succeeded extremely well in producing a coalition of forces that see benefits in this mode of climate change policy (Paterson 2010). This is a coalition that includes a broad swathe of mainstream environmental NGOs (most of whom now give qualified support to the ETS), financiers, project developers, lawyers, and the regulated firms who prefer it to any other policy design. The traders are now, for example, organised in the *Carbon Markets and Investors Association*, that lobbies openly for more aggressive emissions reductions. It can be regarded, in effect, as a hegemonic coalition in European climate policy, and has displaced the interests of the main interests who attempt still to stall climate policy *per se*. It is in this that the ETS can be regarded as a considerable success, as no other overall policy design on climate change can be said to have done this. This of course renders it highly problematic, given it is at its core an uneasy alliance of environmentalism and finance. But critics are in our view unwise to simply ignore its political benefits; the challenge then is to mobilise this coalition for emissions reductions rather than just financial benefits.

Offsets

Broadly speaking there are two types of offset market in the carbon economy: one which is overseen by the UN and states which enables countries to comply with their emissions reductions commitments in the Kyoto Protocol (the compliance market) and another which is purely organised by and for private actors, though with the increasing involvement of civil society organisations (the voluntary market).

The Clean Development Mechanism, often described as the ‚Kyoto surprise' (Werksman 1998), has also been considerably more successful in terms of levels of finance mobilised and projects approved than its architects or critics had imagined. Its future hangs somewhat in the balance, however, amid ongoing uncertainty in the climate change negotiations over the future of the UNFCCC process and Kyoto Protocol that established the CDM. The CDM is a key feature of the Kyoto Protocol's flexibility measures, insisted on by the US in particular as a precondition for their acceptance of a new agreement. It operates a system whereby industrialised countries can offset their emissions by funding certified emissions reduction projects in developing countries. Politically this satisfied domestic capital in leading states such as the US that emissions reductions could be pursued in the lowest cost way, while also providing a way to enrol rapidly industrialising countries in efforts to reduce emissions within the global climate regime.

The logic of efficiency dictates that in an allegedly open and globalised economy it makes sense to reduce emissions wherever it is cheapest to do so. If it is cheaper to reduce a tonne of carbon in India than in the UK, a UK-based investor should be entitled to pay for emissions reductions in India. This is the logic of offsets both in compliance and voluntary markets. The ‚abstraction' of emissions reductions from the sites and social processes in which they are embedded has predictably attracted a great deal of criticism. The Centre for Science and Environment's Anil Agarwal put it the following way:

> Is one tonne of a greenhouse gas produced by a New Yorker or a Londoner equal to a tonne of the same gas produced by a peasant in Guatemala, Chad or Bangladesh? The simple, moral answer is 'no'. The first tonne is the result of luxury. The second tonne of basic survival. Both of them go into the atmosphere. But one needs to be controlled and the other needs to be supported (Agarwal 2000).

Yet for the purposes of trading in offsets, they are the same. Because the value of Certified Emissions Reductions (CERs – the credit, and thus commodity, created in the CDM system) materialises through the medium of a project, a series of procedures have to be undertaken to ‚fix' carbon in particular places for specified amounts of time (Bumpus 2009). This is true at least for first level commodification of carbon. Secondary buying and selling of CERs, however, is not about their intrinsic or use value, but their exchange value determined by market price. Though the process of commodifying hypothetical emissions seems to rely on an abstraction, as Bridge notes (2008), commodification is a creative process that requires the construction of an extensive infrastructure for the production, evaluation and realisation of value including traders, verifiers, accountants and brokers: the ‚market makers' (Newell 2009). The day to day functioning of this aspect of the CDM requires a range of actors to perform distinct governance roles: identifying and developing projects, getting them approved by Designated National Authorities in government ministries and the UN CDM Executive Board and then monitoring and verifying that the emissions reductions have indeed taken place carried out by Designated Operating Entities. Carbon traders and brokers are frequently critical of the slow pace of approval and the hurdles they have to overcome to get projects and methodologies approved and CERs issued (e.g. Dornau 2008). The high transaction costs serve as a deterrent to many potential (and actual) market participants. Yet value is also extracted by actors benefitting from asymmetries of information to capture and add value. As Bridge argues (2008) ‚carbon finance is part of a more general process of expanding value by circulating capital in its money form, rather than in the production of material commodities: as a financialisation of nature', through what Marx called ,fictitious capital' (Marx 1967). As a consequence, ,through its financialisation, the real commodity – carbon – produced and unproduced – is now integral to the multidimensional web of capitalist technology and social organisation' (Smith 2006: 29). It is this financialisation that has allowed financial centres, in particular the city of London, to further consolidate their power by serving as a central hub in the circulation of capital associated with the CDM.

Besides the abstraction from the social value of the emissions noted above, there is another dimension of uneven development that potentially occurs through this exchange. This is the claim that the CDM exacerbates (or reproduces) uneven development in the form of ,climate colonialism'. The key dynamic is that CDM investment potentially locks developing countries into patterns of carbon emissions that impede their development, while industrialised countries get to use offsets to continue their carbon-intensive development paths. Even senior government officials within the climate change negotiations, such as the ,father of Kyoto' Ambassador Raúl Estrada-Oyuela, noted at the time of the Kyoto agreement: ,My reservation was that the CDM is considered a form of joint implementation but I don't understand how a commitment can be jointly implemented if only one of the parties involved is committed to limit emissions and the other party is free from a qualitative point of view. Such disparity has been at the root of every colonisation since the time of the Greeks' (1998).

The functioning of the CDM to date its revealing of its origins as a neoliberal experiment. Capital flows in the CDM have largely mirrored flows of FDI in the developing world with China, India and Brazil the three largest recipients while sub-Saharan Africa continues to attract less than 2% of CDM projects. State and corporate elites in those countries have seen in the CDM the opportunity to extract rents and assert their power to shape the terms of investor engagement to their advantage through taxation schemes that differentiate between project types on the basis of whether they serve national priorities and by concluding Emissions Reductions Purchase Agreements which split the CERs released for projects between state and private actors (Schroeder 2009; Newell et al 2009). The tendency to pursue low-hanging fruit strategies where large volumes of CERs can be generated from modest, incremental and low-cost changes to business as usual production has been affirmed. In the CDM market, 70% of CERs in the first year and half were issued for abating gases other than CO_2 in particular the destruction of industrial gases used in refrigeration (Paulsson 2009). Because CERs are weighted according to the global warming potential of a gas, this created incentives to target projects aimed at removing gases such as HFCs rather than more difficult and longer term investments in renewable energy for example. While projects are also obliged to show how they contribute to sustainable development, such contributions are harder to quantify and therefore are not rewarded financially or with CERs and thus often get neglected (Olson 2007). The drive to capture value at minimum cost makes it rational for sponge iron facilities, aluminium plants, cement factories and municipal waste dumps to make minor changes to production processes in order to get credit (and climate finance) for emissions reductions. In the worst case, this can be a subsidy to polluting activities which affect the poor most seriously, where, for example, waste sites are given a new lease of life by receiving payment for capturing and burning methane, simultaneously undermining campaigns for their closure (Lohmann 2005; Newell 2009).

In another parallel with financial markets in general, there is significant potential for ‚climate fraud‘ in the CDM – double-counting of the same project by selling emissions reductions to more than one buyer simultaneously, and dubious ‚additionality‘ (required to show that the emissions would not have been achieved using existing technologies and finance or are required by existing regulation) as critics of carbon trading have been quick to point out (Lohmann 2006; Böhm and Dabhi 2009). In a context of a financial crisis started by the over-allocation of sub-prime mortgages and the rise of toxic debt, activists have been quick to dub carbon credits as ‚sub-prime‘ and ‚toxic‘ carbon amid fears of the consequences of placing the fate of the climate in the hands of the very people that brought about one of the worst financial crises on record (FoE 2009).

Interestingly, in an attempt to preserve the legitimacy of offsets as a valid means of addressing climate change, beneficiaries and participants in offset markets have sought to respond to these criticisms by establishing forms of private voluntary regulation and certification. These include the *Gold Standard* which only accredits projects on renewable energy to demonstrate that not all investors are only interested in the cheapest low-hanging fruit options and the *Voluntary Carbon Standard* and the *Offset Quality Initiative* which seek to impose higher levels of quality control and assurance. To capture value from projects which bring higher sustainable development benefits, initiatives such as the *Climate, Community and Biodiversity Standards, Social Carbon* and *Plan Vivo* have been developed in relation to forestry projects. As a conscious attempt to maintain the credibility of offset markets as responses that represent meaningful emissions reduction opportunities, as well as new sites of accumulation, these private schemes for private actors potentially deliver public benefits. It is also apparent that there is more governance of carbon markets, including offsets, than critical and conventional accounts give credit for (Newell and Paterson 2010). Abyd Karmali, Managing Director, Global Head of Carbon Markets, Merrill Lynch puts it succinctly:

Those who assume that the carbon market is purely a private market miss the point that the entire market is a creation of government policy. Moreover, it is important to realise that, to flourish, carbon markets need a strong regulator and approach to governance. This means, for example, that the emission reduction targets must be ratcheted down over time, rules about eligibility of carbon credits must be clear etc. Also, carbon markets need to work in concert with other policies and measures since not even the most ardent market proponents are under any illusion that markets alone will solve the problem.[6]

It is also notable that there is some evidence of what the World Bank calls a ‚flight to quality‘ in voluntary offset markets which increasingly emulate the procedures and *modus operandi* of the compliance market in terms of the use of project design documents, third party verification, and the more stringent of the standards mentioned above. Voluntary offsets provide services to firms wanting to project their green or socially responsible credentials as well as appeasing the guilt wealthy individuals feel when they fly for example. Offsets allow these actors to claim carbon neutrality and therefore no net contribution to the problem of climate change. Such claims have been subject to ridicule (Smith 2007), as well as being charged with sapping the impetus and pressure on polluters to bring down their own emissions by displacing that responsibility elsewhere. Politically, nevertheless, they have provided a ‚safe‘ way of drawing firms into debates about their responsibilities to act on climate change. The same might be said of governments that chose to stay out of Kyoto. Firms that see the potential in carbon markets in countries that until recently were outside the Kyoto regime such as the US and Australia have provided significant demand for offsets. Offset firms speak confidently of the ability to ‚engineer a shift in political stance‘ of some key players that had been opposed to action, once they see the money to be made.[7]

3   Making sense of the carbon economy

So how are we to make sense of the ways in which capitalism has attempted to render climate change non-threatening and ‚treatable‘ within its existing structures and modes of operation? A critique of the carbon economy that has resulted from climate policy's neo-liberal heritage is readily available. Such a critique in

---

[6] Is carbon trading the most cost-effective way to reduce emissions? ClimateChangeCorp: Climate News for Business http://www.climatechangecorp.com/content.asp?ContentID= 6064. Accessed April 9 2009.
[7] Conversation with Johannes Ehberling, Eco-Securities, Oxford, February 28th 2008.

effect argues that climate policy has been hijacked, and environmentalism co-opted, into a set of approaches that serve capital's drive for constant accumulation well, but which do little to reduce emissions. Empirically, such critiques focus on the ‚climate fraud and carbon colonialism' which carbon markets exhibit (Bachram 2004; Lohmann 2006). Theoretically, it arises out of a critique of commodification which derives from Marx (Lohmann 2010, Castree 2003; Prudham 2009) and Polanyi (Lohmann 2006; Bumpus and Liverman 2008; Paterson 2010), arguing in effect that the privatised form of the commodity is in direct contradiction with the structural requirements of climate change as a global public good.

We have many sympathies with such a critique of carbon markets. However we want to take issue in three ways.

First is that empirically, we are not convinced that the evidence is that watertight that carbon markets cannot, if well-designed and regulated (a big if, admittedly), play a co-ordinating role in shaping the global economy towards decarbonisation. Their potential to provide system-wide incentives for investors, producers and consumers, combined with the fixed caps that at least emissions trading requires, seems to us unwise to dismiss out of hand. The widespread assumption in critiques of marketisation that because it involves commodification, it cannot ‚work' in terms of emissions reductions, is to ignore the contradictions internal to any project within capitalism. But contradictions in the Marxist sense go both ways in this situation; just as for example it explains the inability of capital to realise its objectives fully, it also explains the unintended consequences and complex politics, in this case that the rise of finance has produced system-wide co-ordination possibilities through financial markets that might be used to achieve environmental goals.

Second is that we need to think politically about climate policy, which means in capitalist conditions that the construction of a coalition of forces that can overcome the objections of interests threatened by climate policy is crucial in imagining a political process that might decarbonise the economy. Carbon markets arguably have proved extremely useful in this regard. Clearly, for any political-economy approach, the relationship between the state and capital is central to an enquiry into the ways in which the carbon economy has come into being and evolved and the extent to which states can use markets to address market failures or in the case of climate change what Nicolas Stern described as ‚the world's greatest market failure' (Stern 2007). Given the nature of relationship between the state and capital in conditions of capitalism, the mutual dependencies that exist and the structural power of capital in relation to the state (Gill and Law 1988; Holloway and Picciotto 1978), responses to climate change have to be negotiated with powerful sectors of the economy. Instances where industry have mobilised on a widespread scale to veto policy developments they oppose, often on grounds of carbon leakage accompanied with threats to re-locate, have been highly successful. The use of the Byrd-Hagel resolution in the US to veto its participation in the Kyoto Protocol and the aggressive and intensive campaign by large energy users against the proposed EU carbon tax in 1992 provide clear examples of this (Newell and Paterson 1998). They suggest the dangers of failing to engage powerful actors and fractions of capital willing to depart from opposi-tional positions.

But of course the interests of capital in relation to climate change are not monolithic. They have to be expressed in terms which demonstrate a contribution to the interests of capital in general. Early on in the process, it was relatively straightforward for energy producers to argue that their business served the inter-ests of capital in general because of the close relationship between energy use (largely based on fossil fuel use) and economic growth (Newell and Paterson 1998). But that situation has become significantly more complex since the mid 1990s, with the emergence of various parts of large transnational business (from the energy sector, to insurance, to financiers, to some parts of manufacturing) seeing increasing benefits in emissions reductions, especially if organised through an overarching policy framework organised around emission trading. In this con-text, thinking about the state-capital relation with regard to climate change does not leave us with a singular account of how that politics will play out.

Third is that the process of resistance to carbon markets should be thought of as internal to the politics, not as an add-on extra (Paterson 2009; 2010). Be-yond the strategic attempts to create new alliances with powerful elements within neo-liberalism, such as Greenpeace's attempts to court the insurance industry (Paterson 2001) or the attempts to use investor power to push firms to disclose their emissions described above, a broader set of critics of carbon markets, who are sceptical of capitalism's ability to reconcile its growth requirements with efforts to reduce GHG emissions, have sought to expose the scams and injustices associated with carbon markets. Rather than stand apart from the system, never-theless, the effect of their campaigns and exposés, is to create better functioning markets where advocates and those with a stake in their success are forced to deal with these critiques and demonstrate that they do not diminish their ability to deliver emissions reductions in a profitable way. That is, actually existing carbon markets reflect in a number of ways concerns about climate fraud and carbon colonialism. The virtual exclusion of forests from the CDM to date is one example, the emergence of certification systems in the voluntary markets noted above another. This is in part the answer to the ‚a big if' in the first point above, since it follows that continued critique of and social movement activism against carbon markets is crucial (if ironically so) in maximising the potential of such markets in relation to emissions reductions and global justice.

Theoretically, therefore, we would argue that the politics of carbon economy should be understood in two general senses. One, focused on the politics in the narrow sense, is through a Gramscian notion of hegemony. In this sense, the projection and reinforcement of common sense ideas about the efficiency of markets as vehicles for reducing emissions is also important, as is the creation of a constituency of business actors with an interest in low carbon development and particularly in carbon markets. In our view, there now exists at least an historic bloc in construction which we can imagine sustaining climate change policy focused on increasingly radical policies to reduce GHG emissions over the coming decades. This coalition of forces is transnational in scope, not limited to a number of ‚progressive' environmental ‚leader' states like the UK, Germany or Sweden. Hence even when the US chose to extract itself from the climate change negotiations in 2001, some US businesses sought to develop their own responses that allowed them to capitalise on emerging opportunities in the carbon economy through the Chicago Climate Exchange, for example.

Actors part of this bloc have an interest in ensuring that carbon markets are the preferred and most profitable way of responding to climate change. This means they have to engage in accommodation of critics, working with civil society organisations in standard-setting processes and demonstrating to publics and consumers of their services, that emissions reductions are genuine and can come with social benefits. Despite their promise as bottom-up *laissez-faire* responses to the threat of climate change, carbon markets require elaborate and vast systems of accounting, verification, property rights and robust institutions to work. From a more Polanyian perspective, we might expect further moves to re-embed (carbon) markets in new frameworks of social control to address the problems that have been identified and exposed.

At a deeper level, however, we can see the dynamics of hegemony at work which lend themselves to a (neo) Gramscian analysis (Levy and Newell 2002). Attempts to secure the legitimacy and salience of market based responses by carbon brokers, traders and other beneficiaries of the carbon economy can be seen as attempts to preserve the hegemony of the material base, institutional apparatus and ideational claims of neo-liberalism. The development of private regulation, codes and standards of certification such as the Gold Standard or Voluntary Carbon Standard can be seen as an attempt to deal with activist and public concerns about the social and environmental problems produced by carbon markets while preserving their status as the preferred response to climate change. At the same time, we have seen attempts by activists to engage with fractions of capital that wield considerable power within neo-liberalism, most notably finance. We gave the example of the CDP which seeks to mobilise the power of investors to pressure firms to disclose and reduce their emissions, but could also point to the work of actors such as the Climate Group making the business case for action on climate change and piecing together interesting coalitions between cities, firms and citizens. Many of those involved have shifted their strategic focus from the negotiating halls of the UN as a way of seeking short-term action on climate change and towards those business actors which wield considerable power and whose actions could deliver cuts in emissions which dwarf those that could be achieved through multilateral diplomacy (Newell 2005).

Alongside this Gramscian analysis of the politics of carbon markets and climate policy, we would add an account of the economic dynamics of decarbonisation drawn loosely from regulation theory (Boyer 2004). This approach is particularly useful because it accepts certain basic structural features of capitalism, but highlights the contingency of particular efforts to overcome capitalism's contradictions, and the search that capital is therefore routinely engaged in for relatively stable patterns of growth – a regime of accumulation. In this context, the emergence of climate change can be understood as coming precisely in a moment where there is no stable regime of accumulation in place. There is rather a dominant strategy to create one around financialisation as outlined earlier, but the coherence of that strategy is far from assured. A focus on climate change is also suggestive in terms of drawing our attention to the socio-technical basis of the classic regime of accumulation – fordism – in the energy technologies centred on coal and oil use and the forms of growth (automobiles, mass produced housing, a range of mass consumption items premised on electrification) they were associated with. Given that context, the strategic question for promoters of climate policy is to articulate the various specific policies – promotion of renewable energy, energy efficiency, and so on – in terms of an overall regime of accumulation. Again, the strategy exists in the context of finance-led growth, which is suggestive of the potential for carbon markets to serve as co-ordination devices which create the relatively stable relations between production, consumption and investment characteristic of a regime of accumulation. Once again, such a regime will necessarily contain its own contradictions, but the existence of contradictions is not a reason not to pursue something.

4   Conclusion: Towards Climate Capitalism

Given the dynamics of capitalism and its contemporary neo-liberal variant that we have described in this chapter, it is clear that short or medium-term transitions to a low carbon economy will have to be supported (financially and politically) by powerful fractions of capital with a stake in the success of such a project. Identifying growth opportunities in low carbon investment technologies will require a suite of different policy tools and mechanisms, but ultimately it needs a powerful political constituency to drive it forward in the face of challenge, scepticism and contestation.

Our analysis in the book *Climate Capitalism* identifies a series of scenarios in which financial capital might be brought on board to accelerate the decarbonisation of the economy. Instruments such as the CDP, alliances with bankers and the insurance industry will all be critical to sensitising business to the need to shift investments away from fossil fuels and into renewables and low carbon technologies. There remains a critical role for states and international institutions in setting clear targets, which they clearly failed to do in Copenhagen in December 2009, and creating an enabling environment for such a transition to occur – insisting on disclosure of carbon emissions, offering tax credits and subsidies to emerging renewable enterprises and those firms that deliver substantial emissions cuts.

The extent to which one strategy is adopted over another will also of course depend on a variety of unpredictable factors such as the price and availability of oil, broader geo-political events (such as wars) and the overall health of the economy. But it is clear that whether it occurs through carbon markets or not, strategies to address climate change have to engage with those that exercise power in contemporary capitalism. Identifying viable accumulation strategies and the political constituencies to support and realise them is vital to this enterprise.

References

Aglietta, M. (1999): La globalisation financière. In CEPII (ed.): L'économie mondiale 2000. Paris: La Découverte for the Centre d'études prospectives et d'informations internationales: pp. 52-67.

Altvater, E. (2006): The social and natural environment of fossil capitalism. In: Panitch, Leo; Leys, Colin (eds.): Coming to Terms with Nature Socialist Register 2007. Monmouth: The Merlin Press: pp. 37-60.

Bachram, H. (2004): Climate Fraud and Carbon Colonialism: The New Trade in Greenhouse Gases. Capitalism, Nature, Socialism 15 (4): pp. 10-12.

Benton, T (2000): An ecological historical materialism. In: Gale, Fred; M'Gonigle; Michael (eds.): Nature, Production and Power: Towards an Ecological Political Economy. Cheltenham: Edward Elgar: pp. 83-105.

Böhm, S.; Dabhi, S. (eds.) (2009): Upsetting the Offset: the political economy of carbon markets. Colchester: Mayfly books.

Boyer, R. (2004): Une théorie du capitalisme, est-elle possible. Paris: Odile Jacob.

Boyer, R. (2000): Is a Finance-led growth regime a viable alternative to Fordism? A preliminary analysis. Economy and Society, 29 (1): pp. 111-145.

Boykoff, M.; Randalls, S. (2009): Theorising the carbon economy. Environment and Planning, A Vol. 41: pp. 2299-2304.

Bridge, G. (2008): The new carbon economy: What's new? Notes for a workshop at Christ's College Oxford September 4th 2008.

Brunnengräber, A. (2006): The political economy of the Kyoto Protocol. In: Panitch, Leo; Leys Colin (eds.): Coming to Terms with Nature Socialist Register 2007: pp. 213-231.

Buck, D. (2006): The ecological question: Can capitalism prevail? In: Panitch, Leo; Leys, Colin (eds.): Coming to Terms with Nature Socialist Register 2007. Monmouth: The Merlin Press: pp.60-72.

Bulkeley, H.; Newell, P. (2010): Governing Climate Change. London: Routledge.

Bumpus, Adam; Liverman, Diana (2008): Accumulation by Decarbonization and the Governance of Carbon Offsets. Economic Geography, Vol. 84, No. 2: pp. 127-155.

Capoor, K.; Ambrosi, P. (2009): State and Trends of the Carbon Market 2009. Washington DC: World Bank.

Castells, M. (1996): The Rise of the Network Society. Oxford: Blackwell.

Castree, N. (2003): Commodifying what nature? *Progress in Human Geography* 27, 2: pp. 273-292.

Castree, N. (2006): From neo-liberalism to neo-liberalisation: Consolidations, confusions and necessary illusions. Environment and Planning, A Vol. 38: pp. 1-6.

Dalby, S.; Paterson, M (2009): Over a Barrel: Cultural Political Economy and ‚Oil Imperialism'. In: Debris, François; Lacy, Mark (eds.): The Geopolitics of American Insecurity: on Terror, Power, and Foreign Policy. London: Routledge: pp. 181-197.

Dornau, R. (2008): Defending the Integrity of the CDM. In: Carnahan, K, (ed.): Greenhouse Gas Market 2008: Piecing Together a Comprehensive International Agreement for a Truly Global Carbon Market. Geneva: International Emissions Trading Association: pp. 77-81.

Durban Declaration, (2004): Climate Justice Now! The Durban Declaration on Carbon Trading. Signed October, 10th 2004, Glenmore Centre, Durban, South Africa.

Estrada, R. (1998): First approaches and unanswered questions. In: Issues and Options: The Clean Development Mechanism. New York: UNDP: pp. 23-29.

Frank, T. (2000): One Market Under God. New York: Anchor Books.

Friends of the Earth (2009): Sub-Prime Carbon: Re-thinking the world's largest new derivatives market. Washington: FoE.

Gamble, A. (1988): The free economy and the strong state: the politics of Thatcherism. London: Macmillan.

Gilbertson, T.; Reyes, O. (2009): Carbon Trading: how it works and why it fails, Critical currents no. 7, November. Uppsala Dag Hammarskjöld Foundation.

Gill, S.; Law, D. (1988): The Global Political Economy. John Hopkins University Press.

Gould, K.; Pellow, D.; Schnaiberg, A. (2008): The Treadmill of Production: Injustice & Unsustainability in the Global Economy. Boulder, CO: Paradigm Publishers.

Harvey, D. (2005): A Brief History of Neoliberalism. Oxford: OUP.

Helleiner, E. (1994): States and the Reemergence of Global Finance: From Bretton Woods to the 1990s. Ithaca NY: Cornell University Press.

Heynen, N.; McCarthy, J.; Prudham, S.; Robbins, P. (eds.) (2007): Neoliberal Environments: False promises and unnatural consequences. London: Routledge.

Holloway, J.; Picciotto, S. (eds.) (1978): State and Capital: A Marxist Debate. London.

Kovel, J. (2002): The Enemy of Nature: The End of Capitalism or The end of the World. London: Zed books.

Levy, D.; Newell, P. (2002): Business strategy and international environmental governance: Toward a neo-Gramscian synthesis. Global Environmental Politics Vol.3 No.4, November 2002: pp. 84-101.

Leyshon, A.; Thrift, N. (1996): Money/space: geographies of monetary transformation. London: Routledge.

Lohmann, L., (2005): Marketing and making carbon dumps: Commodification, calculation and counter-factuals in climate change mitigation. Science as Culture Vol.14 No.3: pp. 203-235.

Lohmann, L., (2006): Carbon Trading: A Critical Conversation on Climate Change, Privatisation and Power. Development Dialogue No.48, September. Uddevalla Sweden: Mediaprint.

Lohmann, L. (2010): Commodity fetishism in climate science and policy. Presentation to Imperial College London, February.

Marx, K. (1967): Capital Volume III. New York: International publishers.

Marx, K. (1975): The German Ideology Vol. 5. London: Lawrence and Wishart.

Marx, K.; Engels, F. (1978): Manifesto of the Communist Party. Cited from Renton, D. (ed.) (2001): Marx on Globalisation. London: Zed books: p. 28.

Mol, A., (2003): Globalization and Environmental Reform: The Ecological Modernization of the Global Economy. Cambridge MA: MIT Press.

Newell, P. (2005): Climate for Change: Civil society and the Politics of Global Warming. In: Holland, F. (ed.): Global Civil Society Yearbook. London: SAGE.

Newell, P. (2008): The Marketisation of Global Environmental Governance: Manifestations and Implications. In: Park, J.; Conca, K.; Finger, M. (eds.): The Crisis of Global Environmental Governance: Towards a New Political Economy of Sustainability. London: Routledge: pp. 77-96.

Newell, P. (2009): Varieties of CDM Governance: Some Reflections. Journal of Environment and Development, Vol. 18 No. 4: 425-435.

Newell, P.; Jenner, N.; Baker, L. (2009): Governing clean development: A Framework for Analysis. Development Policy Review Vol.27 No.6: pp. 717-741.

Newell, Peter; Paterson, Matthew (1998): A climate for business: Global warming, the state and capital. Review of International Political Economy Vol.5 No.4 Winter: pp. 679-704.

Newell, P.; Paterson, M. (2009): The politics of the carbon economy. In: Boykoff, M. (ed.): The Politics of Climate Change: A Survey. London: Routledge: pp. 80-99.

Newell, P.; Paterson, M. (2010): Climate Capitalism: Global Warming and the Transformation of the Global Economy. Cambridge: CUP.

O'Connor, M. (ed.) (1994): Is Capitalism Sustainable? Political Economy and the Politics of Ecology. New York: Guilford Press.

Olsen, K. H. (2007): The Clean Development Mechanism's Contribution to Sustainable Development: A Review of the Literature. Climatic Change 84: pp. 59-73.

Paterson, M. (2001): Risky Business: Insurance Companies in Global Warming Politics. Global Environmental Politics 1 (3): pp. 18-42.

Paterson, M. (2009): Resistance makes carbon markets.. In: Böhm, Steffen; Dabhi, Siddhartha (eds.): Upsetting the Offset: the political economy of carbon markets. Colchester: MayFlyBooks: pp. 244-254.

Paterson, M. (2010): What are carbon markets for? Politics and the development of climate policy. paper under review.

Paulsson, E. (2009): A review of the CDM literature: from fine-tuning to critical scrutiny?. International Environmental Agreements: Politics, Law and Economics, Springer, vol. 9(1): pp. 63-80.

Philips, L. (2009): EU emissions trading an „open door" for crime, Europol says. EU Observer. Available at http://euobserver.com/9/29132, accessed April, 7th 2010.

Pollard, J.; Oldfield, J.; Randalls, S.; Thornes, J.E. (2008): Firm finances, weather derivatives, and geography. Geoforum, 39: pp. 616-624.

Prudham, S. (2009): Commodification. In: Castree, N,; Demeritt, D.; Liverman, D.; Rhoads, B. (eds.): A Companion to Environmental Geography. Oxford: Blackwell.

Pryke, M. (2007): Geomoney: An option on frost, going long on clouds. Geoforum, 38: pp. 576-588.

Roberts, T. J.; Grimes, P.; Manale, J. (2003): Social Roots of Global Environmental Change: A World-Systems Analysis of Carbon Dioxide Emissions. Journal of World-Systems Research 9 (2): pp. 277-315.

Roberts, T. J.; Parks, B. (2007): A Climate of Injustice: Global Inequality, North-South Politics and Climate Policy. Cambridge MA: MIT Press.

Sandbag (2010): The Carbon Rich List: The companies profiting from the EU Emissions Trading Scheme. Sandbag, February.

Sandler, B. (1994): Grow or die: Marxist theories of capitalism and the environment. Rethinking Marxism Vol.7: pp.38-57.

Schroeder, M. (2009): Varieties of carbon governance: Utilising the CDM for Chinese priorities. Journal of Environment & Development, Vol. 18, No. 4: pp. 371-394.

Smith, K. (2007): The Carbon Neutral Myth: Offset Indulgences for your Climate Sins. Amsterdam: Carbon Trade Watch.

Smith, N. (2006): Nature as accumulation strategy. In: Panitch, Leo; Leys, Colin (eds.): Coming to Terms with Nature Socialist Register 2007: pp. 16-37.

Stern, N. (2007): The Stern Review: The Economics of Climate Change. Cambridge: CUP.

Storper, M.; Walker, R. (1989): The Capitalist Imperative: Territory, Technology and Industrial Growth. Cambridge: Blackwell publishers.

Streck, C. (2004): New partnerships in global environmental policy: The Clean Development Mechanism. Journal of Environment and Development, Vol.13 No.3, September: pp. 295-322.

Werksman, J. (1998): The Clean Development Mechanism: Unwrapping the Kyoto surprise. Review of European Community and International Environmental Law 7: pp. 147-58.

World Bank (2008): Climate Investment Funds. Washington DC. Available at www.worldbank.org/cifs, accessed February, 4th 2011.

Climate Politics as Investment
An Analysis of the Discourse on Financing Mitigation and Adaptation[1]

Simon Wolf

1 Introduction

Climate politics in the Post-Kyoto-Phase apparently has turned into an issue of finance and investments: Enhancing north-south financial flows was one of the sticking points for agreement at COP 15 in Copenhagen 2009, and the failure to achieve substantial progress one this issue caused strong tensions between developing and developed countries; thereafter, the UN Secretary General launched a High Level Advisory Group on Climate Finance to determine the required levels of finance and the potential of different sources; the World Bank and many national governments have launched new climate finance instruments; and a whole new range of actors engages a discussion on how to meet the climate challenge through scaling up climate finance. This new importance of finance is due to a new attention for climate protection in developing countries, which is explained through the general need for higher emission reductions, the quickly rising emissions in some developing countries, and the comparably cheap abatement opportunities in the global south.

New emphasis on investments

What becomes apparent when looking at many of the recent developments is the new emphasis that is given to raising private investments for climate protection. Given the magnitude of the climate challenge, and the level of climate finance required accordingly, the conclusion is drawn that all available sources of money must be used, and that the vast majority of financial flows will have to come from the private sector. The role of policy and public is to support investors and incentivise investments flows.

---

[1] An earlier version of this paper was presented at the SGIR 7th Pan-European International Relations Conference, 9-11 September 2010, Stockholm.

Much of the climate finance debate and the initial political regulation, in consequence, focus on this objective: Proposals for Public Finance Mechanisms, like the World Bank's Climate Investment Funds (CIF) discuss how to increase the leverage effect of public spending on private investments, by providing guarantees, loans and grants (Maclean et al. 2008); the carbon market discussion has shifted from the capacity of markets to identify cost-efficient emission reductions to their potential to stimulate investments; and while a general consensus exists for the implementation of a forest finance mechanism, the most contentious issue is if and how markets and private investments can contribute to the objective of forest protection.

While many practitioners like government departments, NGOs, and consultancy firms have taken up the issue, the (analytical) social science literature on finance in climate politics almost entirely remains focused on carbon markets, weighing their pros and cons against the background of the Kyoto situation. But while the focus was on emission reductions in developed countries then, and reducing emission in developing countries was only a minor issue delegated to the Clean Development Mechanism (CDM), the situation has dramatically changed in recent years.

This raises the question whether the heavy criticism of the marketisation of climate politics should be revisited, that is, whether the changing context adds new legitimation to the use of market and investment instruments; or, to the contrary, whether the urgency that accompanies the calls for raising climate finance and incentivising private investments relies on a narrow problem framing, that rashly neglects alternative approaches and solutions.

To answer these questions, this paper analyses the climate finance debate, describing how the focus on raising investments has emerged. It argues that these developments can only be understood against the background of an economic rationality in climate politics that makes climate change legible and manageable as an economic problem (section 2). Much of the current climate finance debate is organised according to this logic, calling for cost-effective mitigation and incentivising private investment, thereby taking an active role in shaping needs and priorities rather then aiming to generate funds for politically defined objectives (section 3). This affects on proposals like Public Finance mechanisms and the REDD mechanism (section 4). Reflecting and questioning the dominant climate finance logic through alternative framings shows that traditional forms of regulation could be more appropriate (section 5).

Some remarks on theory and methodology

The paper presents and discusses some of the (preliminary) findings of a PhD project that analyses the climate finance discourse. While the issues discussed here could be of crucial interest from a Political Economy perspective, as the transformation to low-carbon societies aims at the very heart of the organisation of capitalist societies, or from a governance perspective, as this reorganisation produces new forms of interaction between governments and market actors, the paper focuses on the discursive construction of the climate finance field.

Starting from a poststructuralist ontology, the research strategy combines archeological and genealogical methods: While the *archeological* dimension of the analysis refers to the form of a dominant rationality, the *genealogical* dimension refers to the formation of this rationality based on societal practices and changes to these practices (Foucault 1976/1998).

In applying this research strategy, the project is able to overcome the shortcomings of two approaches to political processes beyond the governmental level that are used for analysing climate politics. Contrary to *constructivist approaches* that are mainly occupied with collectively held ideas and understandings on social life, but miss a „more specific understanding of who the relevant actors are, what they want, and what the content of social structures might be" (Okereke et al. 2009), the genealogical dimension of the research seeks to account for both the role of actors and structures in establishing a particular perspective. And against governance approaches, which focus on the interaction of state and non-state actors but lack an explanation for what becomes an object of governance and in what way, a poststructuralist approach accounts for the role of power in those processes by challenging the rationality that informs political regulation and sets the „ground rules for governance" (Jessop 1999).

2 The Economics of climate change

Economic considerations, perspectives and strategies are all but new in climate politics: As early as 1992, William Cline published his book „The Economics of Global Warming", and ever since, the economic consequences of climate change have been a topic within academia (Cline 1992); business organisations were an important lobby group from the early days of international climate negotiations and achieved to prevent stronger regulation in many cases (Levy and Egan 2003); the UN Framework Convention on Climate Change, agreed on in 1992, states that „policies and measures to deal with climate change should be cost-effective so as to ensure global benefits at the lowest possible cost" (UNFCCC

1992); five years later, the Kyoto Protocol introduced three flexible mechanisms that aim at enhancing the cost-effectiveness of climate protection through the efficiency of markets; and the development of the global carbon market in the following years opened a lucrative playing field for project developers, investment companies, and consultancy firms.

2.1 Reframing climate change as an economic problem

Against this background, it seems disputable to claim that the Stern Review *The Economics of Climate Change* (Stern 2006), and its intensive discussion thereafter, reframed climate change as an economic problem. However, the report marks a crucial turning point in the climate discourse, by establishing a certain economic rationality in climate politics.

It certainly was not the Stern Review alone that brought about these changes, but there can be no doubt regarding its massive influence: The report resonated heavily in mainstream and expert media, and its findings are an important argument in the debate until today. Nicholas Stern, not particularly an expert in climate politics until then, is today „the global authority on climate change" (The Guardian). The question arises, however, what caused this massive resonance, as Stern and his team do by no means present an entirely new perspective. Rather, they reflect and reinforce a changing understanding of the climate change problem.

To adequately understand the importance of the Stern Review, we have to distinguish between its main message that was widely taken up in the public and heatedly discussed in expert circles, and the particular approach to climate politics the report establishes.

Stern and his team compare the global macroeconomic costs of climate change, with and without mitigation policies, outlined for time horizons of 50, 100 and 200 years (Stern 2006). The core message is that the economic consequences of uncontrolled climate change far outweigh the costs of limiting climate change to an acceptable degree: mitigating climate change, therefore, is perfectly rational from an economic point of view.

It is this message that was strongly taken up and almost entirely welcomed in the public debate. But the constraints to the – scientific – relevance of the argument become apparent when looking at the critical debate among economists that followed the publication of the review (Wolf 2009). While many objections were made regarding the chosen methodology, the debate culminated around the choice of the discount rate.[2]

The Stern Report chooses a significantly lower rate than comparable studies, and in consequence, today's climate protection efforts have a higher value in the Stern Report than in most other models (Weitzmann 2007, Spash 2007). This is heavily criticised by other economists who insist that a higher discount rate would have caused far less clear conclusions about the costs and benefits of climate protection.[3]

What is important here is that the choice of the discount rate and other important parameters in economic modelling cannot be justified through scientific reasoning but rather reflect a whole range of assumptions as much as worldviews and beliefs (Ackermann 2007): „Modelling over many decades, regions and possible outcomes demands that we make distributional and ethical judgements systematically and explicitly" (Stern 2006: 143).[4]

The main message of the Stern Review, accordingly, was not purely a scientific argument, but rather a political one. Stern himself supports this point of view three years later in his book *Blueprint for a safer planet*, suggesting that the estimates are not exact forecasts but fulfilled an important role in rising the awareness for the urgency of climate politics (Stern 2009).

Additionally, the argument is of limited value from a policy making perspective. Even taking for granted that global climate protection is economically rational in the middle and long run, the situation can be very different from a national policy perspective: Emission reductions of an individual country do not limit the effects of climate change for this same country, and it is by no means certain that the effects of climate change will be economically negative in every case: Rising

---

[2] This factor, put very simplistic, diminishes the value of present investment over time, as a) money loses its value through inflation, and b) future generations are expected to be wealthier through economic growth. For more on discounting in the Stern Report, see Spash 2007; on discounting in general, see Price 1993.

[3] They also accuse Stern and his team of only taking into consideration the most pessimistic forecasts of climate damages, which further elevates the value of climate protection efforts. This argument is countered by Arrow (2007). He agrees that Stern's results differ considerably from other studies. But, even choosing a much higher discount rate (up to 8.5 per cent following his calculations), climate protection measures would still be economically rational (Arrow 2007). Finally, Schneider (2008) hints to another inconsistency within the „Economics of Climate Change": „It is unacceptable to compare future costs to the present scale of the economy, [...] since projected growth rates of the economy swamp all mitigation and adaptation costs typically found in the literature" (Schneider 2008: 3).

[4] This makes Yohe and Tol (2008) ask whether it is necessary to model the costs and benefits of climate protection, as „economic arguments can be made without resorting to dodgy modelling or peculiar assumptions. Taxing greenhouse gas emissions now makes perfect economic sense" (Yohe and Tol 2008: 237).

temperatures improve the economic conditions for some sectors; adaptation policies can have the effect of an economic stimulus; and the growing demand for
climate proof technologies opens up export markets for some countries.

The Stern Review, then, „is a standard piece of normative welfare economics which asks the question: assuming a benevolent global player, what would
you do?" (Interview Simon Dietz). A benevolent global player, so he existed,
would probably make a strong effort to limit climate change, but turn down economic arguments for much stronger normative motivations. Policy-makers, however, have to take the general decision for climate protection without relying on
cost-benefit considerations.

2.2   And turning climate politics into an economic challenge

Nonetheless, the general argument of the Stern Review had great effect on policy
debates by supporting those who want to make the case for cuts in emissions
(Interview Christopher Taylor). This is by no means accidental. The objective of
the report was not only to put as much authority behind the Economics of Climate Change as possible by making it an official government report, commis-
sioned by the then Chancellor of the Exchequer in the UK Gordon Brown and
written by a former World Bank chief economist; but to make a strong case for
the economic dimensions of climate change, and to bring finance ministries into
the game. The review „shifted the debate away from polar bears and unseasonal
summers, and reframed it in the cold hard language of the balance sheet" (The
Guardian, 30 March 2009).

In consequence, and all objections to the approach taken by Stern and his
team notwithstanding, the report was successful in establishing a global economic perspective on climate change, that allows to frame consequences of climate
change and benefits of climate protection in monetary terms.

To achieve this, the report reduces the complexity of climate change consequences to find a common denominator for costs and benefits, and to make them
fit within economic models (Spash 2007). This involves the valuation of ecosystems and livelihoods and implies, inter alia, the problematic equation of monetary losses in different world regions: what means less luxury consumption to
some, may threaten the basic needs of others.

The use of global aggregate values leaves disparities in income distribution
unconsidered, or at least cannot serve to problematise them. The only requirement in the Stern Review regarding incomes is that „future generations should
have the right to a standard of living no lower than the current ones" (Stern 2006:

42). Future developments, which leave the poor poor and make the rich richer, are fully consistent with this premise.

To be fair with Stern: He is among the most progressive in the economists camp, in refusing high discount rates to play down future consequences of climate change, and even critics acknowledge the sophisticated methodology in shifting „from a single-discipline focus cost-benefit analysis to a new inter-disciplinary and multi-disciplinary risk analysis" (Barker 2008: 174). Nevertheless, after the success of the Stern Review, it is widely accepted that climate change is legible through economic analysis and modelling.

What the Stern Review combines, then, is a new seriousness for the need to act – after all, it is global wellbeing that is at stake if climate change is not limited to an acceptable degree – with a strong sense of opportunity. While governments and International Organisations alike welcomed the message that „we must act now", investors praised the reports emphasis on the compatibility of climate protection and economic growth, and the need for enhancing the scope of markets as „huge business opportunity".[5]

If the European emissions trading scheme was practical because it „created a terminology that business could understand" (Hamilton 2008), the Stern Review delivered the textbook for an economic approach to global climate policy, that strongly resonates in climate politics and is an important basis for the climate finance debate.

3   The new climate finance discourse

The climate finance field is undergoing a process of rapid transformation, and currently consists of different arenas that partially interact with each other. A great number of proposals aims at reforming the finance architecture under the UNFCCC (Müller and Gomez-Echeverri 2009), seeks to differentiate responsibilities for the scaling up of climate finance (Climate Works 2009), or to identify new sources for mitigation and adaptation finance, such as levies on air and maritime travel, a tax on emissions trading schemes, or the withdrawal and auctioning of emission permits (UNFCCC 2007, Parker, Brown et al. 2009).

What is of utmost importance here is the changing approach to climate finance. As long as the focus was on adaptation, the terminology was one of justice, and the climate funds under the UNFCCC were seen as a form of compensation for the consequences of climate change in developing countries (Dietz

---

[5] See „Expert reaction to climate change": http://news.bbc.co.uk/1/hi/business/6098612.stm

2006).[6] Consequently, it was hardly ever questioned if the financial transfers for adaptation should be provided as non-conditional payments from public sources.

3.1 Dominant logics in the new climate finance discourse

This changes with the growing attention for the need to reduce emissions in developing countries. Three main reasons for this need can be found in the literature. *First* is a growing attention for the need to reduce emissions on a global scale.[7] A *second* important reason are the growing emissions in particular in fast emerging economies such as China and India, and it is *third* the availability of comparably cheap emission reductions that turns the spotlight to other developing countries. This gives rise to a new climate finance discourse that partly feeds into the debates and proposals for north-south financial flows within the UN-FCCC, but partly occurs beyond this arena as a general change in the logic or rationality of climate politics as well.

Estimating the costs of climate protection

The common starting point of the climate finance debate in its current form is the level of finance that is needed for mitigation and adaptation, either on a global scale or with a focus on developing countries. Such estimates were first put forward for adaptation needs in developing countries by the World Bank in 2006 and the NGO Oxfam in 2007, estimating that US$ 10-40 billion (World Bank 2006) or „at least US $50bn each year" would be necessary in adaptation finance (Raworth 2007).

While these numbers differ from study to study and generally tend to grow over the years – more recently, World Bank and Oxfam suggest that developing countries need US $ 75-100 billion annually by 2020 for adaptation (World Bank 2009, Gore 2010) – the perspective was soon widened to estimate the financial need for mitigation as well.

---

[6] In political terms, the most important step was the installation of several adaptation funds at the Marrakesh COP in 2001 (Dietz 2006): The Special Climate Change Fund (SCCF) and the Least Developed Countries Fund (LDC-Fund) depend on voluntary contributions by developed countries, the Adaptation Fund is financed through a levy on the CDM, and therefore could only start operating with the entry into force of the Kyoto Protocol. Financial contributions to these funds, however, have remained scarce over the years.

[7] The IPCC's fourth assessment report emphasised that previous efforts were not sufficient for preventing dangerous climate change, and the G8 meeting in Gleneagles in 2005 and the Nobel Price for the IPCC and Al Gore helped moving climate change to the centre of the (international) political agenda (Egner 2007, Luks 2008).

Comparability of these estimates is even more limited as most studies calculate the costs of mitigation as a fraction of global GDP, what adds additional uncertainty factors, like the rate of economic growth. The Stern Review, for instance, calculates the annual costs of limiting global warming to 2 degrees somewhere in the range of -1 per cent (net gains) to 3.5 per cent of GDP. This range „reflects a number of factors, including the pace of technological innovation and the efficiency with which policy is applied across the globe: the faster the innovation and the greater the efficiency, the lower the cost" (Stern 2006: xiv).

What is more important than the high level of uncertainty of these „rough top-down modelling exercises" (European Commission 2009: 4) is the degree to which these studies create a particular perspective on climate change mitigation and, thereby, their own object. In that sense, the climate finance debate takes an active role in constituting needs and opportunities. This „*performativity of economics*" (Mac Kenzie 2007) becomes most apparent where the need for climate finance is derived from an economic analysis of different options.

Priority setting along cost effectiveness

Of fundamental importance in this regard is the *Global Greenhouse Gas Abatement Cost Curve* elaborated by the global consultancy *Mc Kinsey*, that gains similar attention to the Stern Review in current climate politics, and in particular in the climate finance debate. The „Pathways to a low carbon economy" report departs from the insight that mitigation efforts „close to the full potential" are necessary to remain within 2-degrees global warming (Mc Kinsey 2009),[8] and compares the potential of different sectors and world regions to contribute to this objective.

The authors consider four potential ways of emission reductions: Energy efficiency, low-carbon energy supply, terrestrial carbon (forestry and agriculture), and behaviour changes, and conclude that two thirds of global emission reductions should take place in developing countries. Two main reasons are given for these comparative cost advantages. First, developing countries have the opportunity to „leapfrog", by leaving out environmental harmful stages of development and jumping directly into more sustainable levels. The second extensive and low price abatement opportunity can be found in the forestry sector, through afforestation and avoided deforestation.

The result depends heavily on a set of assumptions and normative considerations that reduce the complexity of abatement opportunities to abstract catego-

---

[8] The study was prepared by McKinsey, resulting from a common initiative with Think Tanks, companies, and NGOs, among others the Carbon Trust, Climate Works, Shell, Vattenfall, Volvo, and the WWF.

ries: As the *Cost Curve* aims at the most efficient reduction opportunities, it makes no difference where these reductions take place; there is no need for a global adjustment of *per capita* emissions in the long run;[9] and the authors only consider mitigation options „that do not affect the lifestyles of individuals" (Mc Kinsey 2009: 36).

This last premise is the main reason for the limited abatement potential through behaviour changes in industrialised countries: „Changing behaviour is difficult [...] and there is a high degree of uncertainty in these estimates" (ibid.: 29). The authors identify a much larger abatement potential in the forestry and agriculture sector in developing countries, though „educating and mobilizing billions of farmers around the world to change their daily practices is similarly challenging" (ibid. 34).

The analysis by Mc Kinsey is strongly taken up in the climate finance debate. A strategy paper by the think tanks *Climate Works* and *European Climate Foundation* suggest that of the 17 Gt of emissions that will have to be reduced until 2020, „about 12 Gt of abatement will be physically located in developing countries" (Project Catalyst 2009). According to the *Little Climate Finance Book*, „Of the 17 billion tonnes of emissions reductions required in 2020, 70% is achievable in developing countries" (Parker, Brown et al. 2009: 18). Embarking on the argument for cost-effective forest mitigation made by Stern, Mc Kinsey and others, Bettelheim suggests that „near half of the mitigation actions available in the period to 2020 consists of reducing deforestation and improving agricultural practices in the tropics and sub-tropics", because „developed countries face severe limitations on the cost-effectiveness of mitigation actions they can take by 2020" (Bettelheim 2009: 90).

Equity as finance

These and similar arguments that justify the urgency of emission reductions in developing countries directly from a comparison of the relative costs are ubiquitous in the current climate finance debate. Two things are particularly remarkable here. First, the fact that the economic foundation of the argument is not present in many cases shows how deeply economic logics have penetrated climate politics. Many argue that deeper emission reductions are „achievable" in developing countries, while developed countries face „severe limitations" in this regard, transforming financial constraints to mitigation into absolute ones.

---

[9] The calculations for the year 2030 result in *per capita* emissions of 7.7 tons CO2 in industrialised countries; populations in China and India would emit 3.7 tons, and 1.9 tons in developing countries with a significant share of forestry.

Second, this logic changes one of the normative fundaments of climate politics. The basic rationality of the Kyoto Protocol was that the main polluters will have to reduce emissions in the first place, due to historical responsibility as much as economic capacity. However, „unavailability of plentiful cost-effective reductions in the developed world challenges the assumptions and dynamics underlying the Kyoto Protocol and the European emissions trading scheme. Both focus overwhelmingly on forcing dramatic and rapid changes to the energy and industrial infrastructure of the developed world – an approach that was based on historical responsibility and fairness. Unfortunately, what may have seemed equitable and fitting is neither economically achievable nor environmentally sensible" (Bettelheim 2009: 91). Similarly, Pendleton and Retallack (2009) in a paper for the London based think tank IPPR argue that „in a world of limited finance, reductions arguably [must] be undertaken wherever they can be made for the lowest cost", and offer an alternative perspective on burden sharing: „Since emissions reductions in developed countries are insufficient to solve the climate problem, [...] the principles of responsibility and capability might more productively be applied to the financing of global reductions" (Pendleton and Retallack 2009).

Public and private money in climate finance

One of the most contentious issues in the climate finance debate is the share of public and private money for mitigation and adaptation in developing countries. It is here where the split between the old and new climate finance discourse becomes most apparent.

In the context of UNFCCC negotiations, developing countries repeatedly call for public finance additional to existing ODA flows.[10] This is in particular critical when it comes to adaptation finance: „It is the world's poorest and most vulnerable people – on the front line of the climate crisis – that adaptation finance must reach. The interventions needed [...] will not attract investment from the private sector, since they do not generate internal returns"(Gore 2010: 3).[11] For the same reasons, many developing countries reject counting payments for adaptation as ODA, as „climate finance is not aid. It is not an act of charity, or an

---

[10] Public does not necessarily mean that the money does have to come from national budgets in developed countries: Rather, a whole set of so called innovative sources have been suggested, including international taxes and levies, or intermediary institutions that sell emission allowances from mitigation activities and feed the revenues back into a publicly managed fund.

[11] An argument that has long been backed from an economic point of view as well, as adaptation policies often secure public goods, which does not usually happen along profit maximising interests: „In many cases, market forces are unlikely to lead to efficient adaptation" (Stern 2006, Fankhauser 2006).

expression of solidarity with poor countries, but a legal obligation under the UNFCCC" (ibid. 3).

Much of the current climate finance debate beyond the UNFCCC, however, takes another direction. Instead of framing the debate in terms of legal rights and obligations, it starts from the economic analysis presented above: Comparing the magnitude of finance that is needed for mitigation and adaptation to the current financial flows for climate protection and clean development, the conclusion is drawn that all available sources of money must be used, and the bulk of financial flows will have to come from the private sectors.

Focusing on how to raise the required sums, much of the climate finance debate does not discriminate between different sources of money: „By purchasing offsets to meet their domestic targets, developed country emitters, and their consumers, workers, and shareholders, ultimately finance emissions reductions in developing countries. From the viewpoint of developed countries, these transfers are just as much an expenditure of societal resources as ODA and other public financing mechanisms" (Stewart et al. 2009).

The potentially very different effects of the financial flows are sidelined as the various sources are grouped as *climate finance*, and the „requirement for significantly scaled-up finance and investment in the solutions to climate change" (Hamilton 2009), makes it „inescapable that private as well as public sources must be part of a mitigation finance mix" (Stewart et al 2009: 2). One main success of the CDM in this regard is that it „leveraged more finance into GHG emission-reducing projects in developing countries than any other international mechanism" (Streck 2009: 71).

3.2   From costs to investment in climate politics

Two narratives can be distinguished that focus on the role of private investments for climate protection. Very different starting points notwithstanding, they come to a similar conclusion: The role of governments is essential in incentivising investments and supporting investors.

The „need for private investments" narrative

Against the general need to scale-up climate finance, the „Golden rule of public funding [...] suggests that governments should only support those investments that are economically efficient but not financially viable" (Doornbosch and Knight 2008: 24). „The role of public organizations using valuable taxpayer contributions should be in funding those projects which the market forces will not deem profitable" (Schalatek 2009: 22).

But public spending should not only fill the gaps left by the private sector. The role of public funding „is essential to generate the enabling environment for private sector financing fast enough to make a difference in current investment decisions" (World Bank 2008a: 3). Developed and developing countries alike have to get their regulatory frameworks right in order to allow for and incentivise investment flows into renewable energy and clean technologies. This form of „investment grade policy [...] needs to tackle all the relevant factors that financiers assess when looking at a deal" (Hamilton 2009).

The need for leveraging private finance flows is particular high in developing countries, as investors face higher risks here. The higher contribution of developing countries to emission reductions means that „these countries will not only need to gain financial support from developed countries to cover incremental costs, but also to create a set of coherent policies and regulations that play their part in mobilising private capital" (Project Catalyst 2009).

Providing regulatory certainty, however, is only one of the core tasks for governments in order to unlock private investments. „Because only a minority of such investments are inherently financially viable, government-mandated incentives such as carbon pricing, standards, and direct subsidies/feed-in tariffs would be required to generate greater investments in mitigation" (Brinkmann 2009: 135/6, Robins et al 2009).

Given the scarcity of public funds, in particular in the aftermath of the financial crisis, the assumption is that these resources can be best used to leverage private finance, and thereby, profoundly increase the overall level of financial flows (Romani et al. 2009). Carbon markets are expected to play a crucial role here as well, as they „can leverage as much as nine-fold underlying investment in some sectors" (World Bank 2008).

The „investment opportunity" narrative

On the other hand, the new focus on investments in climate politics reflects a general shift „from threat to opportunity" for business and investors (Newell and Patterson 2010). Success stories of investments into renewable energies and clean technologies and future expectations to the growth of these sectors let investors dream of the „single largest wealth creation opportunity in world" (Hamilton 2008). The business opportunities opened up through the carbon markets further strengthen the expectations.

These developments are accompanied by an intentional push to „reframe the debate in terms of investment benefits rather than mitigation costs. [...] It shows that economic growth and job creation in all major economies can be sustained and even increased under ambitious mitigation scenarios". And it shines a light on the potential benefits from reflating the global economy through a global green „New Deal" [...]„ (The Climate Group 2009: 2).

This reflects the fact that the focus of the Stern and Mc Kinsey studies on the most cost effective emissions reductions can only partially inform investment decisions. „It is worth saying: risk and reward is not the same as cost and benefit" (Hamilton 2008). Investors might choose to fund the opportunities with the lowest capital intensity rather than the ones with the lowest cost over time. The challenge therefore is sometimes rather to find „effective ways to incentivise and finance the (sometimes considerable) additional upfront expenditure [...]. It becomes clear that the cheapest abatement opportunities are not always those with the lowest capital spend" (Mc Kinsey 2009: 15).

Additionally, realising investment can be constrained by non-financial factors. In the case of forestry and agriculture, „both costs and investments are relatively low. Here, the implementation challenges are practical rather than economical" (Mc Kinsey 2009: 16) A similar problem is identified in the housing sector: While builders and owners refrain from making investment into energy efficiency because they do not benefit financially from the savings, tenants are equally reluctant when they are not certain to stay long enough to get their return on investment.

In consequence, many voices call for giving more attention to the opportunities of climate protection and addressing the hurdles that investors face. „There are hundreds of billions of dollars of value that we are not accessing. [...] society as a whole is loosing money by investing in an outdated energy infrastructure", because „[we are] so focused on the costs, and on who is gonna bear the costs, and how we are gonna raise the costs of pollution".[12]

Two consequences are drawn from this. On the one hand, „The topic of clean energy investment should be positioned as a strategic, economic, industrial and foreign policy issue [...] We need new analytics in this area, new evidence base bringing forward much more clearly the investment opportunity, and the benefits in short and medium term" (Hamilton 2008). On the other hand, focusing on investments instead of costs is suggested to inform a political agenda that seeks to create conducive investment environments. „An investment perspective could help tease out the relative importance of these four policy planks [market based instruments, policies to support innovation and accelerate technology develop-

---

[12] Bracken Hendricks, Center for American Progress, in the conference The great transformation – Greening the Economy, at Heinrich Böll Foundation Berlin, Friday, May 28th 2010 (own documentation).

ment, regulation [...] to overcome market barriers, forestry] and suggests from where investments for them should come" (Doornbosch and Knight 2008: 5).

4   Tracing the investment logic in climate politics

The climate finance field is under heavy construction, and so it is impossible to determine how a future finance architecture will look like, even more so as it is part of a larger climate politics package that Parties hoped to tie at COP 16 in Cancún in December 2010: While progress was made here on the installation of the Green Fund as an integrative finance mechanism under the UNFCCC, no real steps forward were made in particular on emission reductions pledges, what in turn affects on the availability of climate finance.[13]

However, important developments are currently underway that reflect the finance and investment debate outlined above. This chapter therefore takes a closer look at Public Finance Mechanisms (PFM), that aim at incentivising private investment for climate protection; and the urgency that is given to a REDD mechanism, what reflects the focus on cost-effective emission reductions.

Public Finance Mechanisms: Translating climate policy into investment terms

Taking up the insight that a) private investment is urgently needed for climate protection, but b) its availability is thus far constrained by risks and barriers, PFM aim at leveraging private finance by improving investment conditions. The discussion and implementation of these mechanisms shows how the role of policy making looks like if thought through from an investor perspective.

The result is no less than a process of translation. One important interpreter in this context is a network of British government departments, the London School of Economics, consultancies, banks and investment firms, including some of the first addresses in the world of finance, like Deutsche Bank, HSBC, or the investor group P8. Brought together by Nicholas Stern, they published the report 'Meeting the climate challenge. Using Public Funds to Leverage Private Investment in Developing Countries ' (Romani et al. 2009). The influence of Stern and other participants in the UN Secretary General´s High-level Advisory

---

[13] Greater emission reduction commitments by developed countries will most likely result in an accelerated growth of the carbon markets, and thereby increase finance flows to developing countries. A failure to come to an ambitious agreement, to the contrary, could result in further strengthening processes outside the UNFCCC: some governments fear, for example, that initiatives like the Interim REDD+ Partnership initiated by Norway and France might further gain ground and eventually replace an UNFCCC mechanism (Martone 2010).

Group on Climate Change Financing ensured that the findings of the report fed into the advisory groups´ recommendations presented to COP 16 in December 2010 as well.

Most of the proposals made in the discussion paper embody the fundamental idea that much higher levels of investment are needed to meet the climate challenge. This understanding becomes most apparent when the role of investments for the forestry sector is explained: „The root causes of deforestation are economic in nature, stemming from under-investment in the sustainable production of land-based commodities, including agricultural products and timber. […] Sustainable forestry and agricultural investments can produce attractive returns for private sector investors, but in many countries there is a significant investment gap between project returns and those required by investors mainly because of high risks regarding political stability and land tenure security“ (Romani et al. 2009: 17).

The focus of the recommendations is therefore on leveraging private investments through „risk mitigation and enhancement instruments, in the form of full or partial guarantees and insurances“ (Romani et al. 2009: 10). One instrument that is already used are so called *cornerstone funds*, large commercial funds that often blend public and private money. The public lender can, like in the case of the European Commissions Global Energy Efficiency and Renewable Energy Fund (GEEREF), partially or completely waive its right for returns in order to enhance those of the private sector. The most important of these mechanisms so far are the World Bank's Climate Investment Funds (CIFs), that dispose of much higher level of finance than the mechanisms under the UNFCCC.

A proposal that sheds light on the understanding of the respective roles of public and private actors was initially put forward by the London Accord, a group of mainly London based financial service companies. *Indexed bonds* identify and simultaneously address two fundamental problems: On the one hand, governments need money to finance the Green New Deal, or, in the context of climate finance, to encourage and support financial flows to the low-carbon economies. On the other hand, investors face „political risks“, as the profitability of investments in renewable energies and clean technologies thus far depends on regulation like emission caps, carbon taxes ore feed-in tariffs. While some of these systems work quite successfully, investors do not like the idea that governments could change or withdraw the regulation, thereby lowering the returns of certain types of projects.

This is where index-bonds come into play. The „simple and somewhat subversive“ idea is to lower the political risk faced by investors through creating an economic risk for governments.[14] If governments borrow money from financial markets (that is, they give out bonds), these bonds are indexed to an indicator like the national emissions level or the carbon price. An investor receives an excess return if the issuing country´s emissions are above the published government target, or the carbon price is lower then predicted. „The bond thus provides a hedge against the risk of the issuing government not delivering on its commitments or targets" (Romani et al. 2009: 13). Not only are governments said to have stronger incentives to deliver on their targets in consequence. The penalty they have to pay gives a guarantee in the language investors understand.

REDD: Complexity, concerns, and great expectations[15]

The discussion on the REDD mechanism emphasises the importance of the dominant finance logic in climate politics, both through the urgency and importance that is given to emission reductions in the forestry sector, and the role that private investment is expected to play here.[16]

The expectation of comparably cheap emission reductions brought forestry on the agenda of northern governments as much as potential investors. The Stern Review and the Eliasch Review on *Financing Global Forests* play an important role in supporting this argument (Eliasch 2008, Bond et al. 2009). REDD offers „high potential at low costs" (Dutschke and Wertz-Kanounnikof 2008), and „these lower cost could allow the international community to meet a more ambitious global stabilisation target" (Eliasch 2008: xxi). REDD, therefore, is seen as a „bridge strategy" that allows to postpone emission reductions in other sectors (Lubowski 2008, Karousakis and Corfee-Morlot 2007).

Beyond massively increasing the importance that is given to REDD, the cost argument affects on proposals for its implementation as well. Many suggest that a carbon market based scheme is best suited to raise forest finance levels by attracting private investments (Union of concerned scientists 2009, Andersen 2008). Additionally, „mandatory markets [...] are often preferred because they

---

[14] http://www.london-accord.co.uk/wiki/index.php/Index-Linked_Carbon_Bonds

[15] Given the uncertainty in which form a forest finance mechanism will be implemented, the paper uses REDD (Reducing Emissions from Deforestation and Degradation) as a general term that includes REDD+ (REDD plus conservation, sustainable management of forests and enhancement of forest carbon stocks).

[16] Forestry is not a new issue in climate politics, it already plays a role in the UNFCCC and the Kyoto Protocol. But while the role of forests projects within the CDM was strongly limited due to the concerns many Parties had at that time (Bäckstrand and Lövbrand 2006), they are expected to play a much greater role in climate protection today, and there is little general objection to a forest finance mechanism (UNFCCC 2009).

would assure long-term, continuous, and predictable flows of finance for REDD projects contrary to voluntary funds" (Alvarado and Wertz-Kanounnikof 2007).

Public spending, accordingly, „must be used strategically to stimulate and complement private investment by helping to provide basic readiness requirements and reinforcing the enabling environment for investment" (Dutschke and Wertz-Kanounnikof 2008), and „to minimize the risk for the private sector in investing and to maximize the returns without hampering socio-economic and other environmental benefits" (Gutman and Cabarle 2010).

Contrary to these great expectation, there is strong evidence for the complexity and uncertainties that a REDD mechanism would have to address, and concerns exist in particular on making forests a subject to carbon markets and investment decisions in particular.

From an environmental perspective, the problem of temporal or spatial leakage is far from being resolved. If forestry credits are used as offsets, leakage would globally lead to rising emissions. These concerns are widespread, as discussions within the UNFCCC show, and a recent study by the OECD and IEA concludes that „Margins of error for changes in carbon emissions are currently too large to support implementation of a market mechanism" (Karousakis and Corfee-Morlot 2007).

On the social side of REDD, many NGOs call for guaranteeing the rights and safety of local communities and indigenous peoples, and are concerned over the implementation of REDD thus far (Martone 2010), and in particular the current rush for REDD. The availability of huge amounts of funding will attract many stakeholders and may have consequences on forest governance beyond or even contrary to the intended effects, „such as elite capture of benefits, potential loss of access to land and lack of voice in decision-making. This is because of the likely scale of the systems envisaged, [...] and the strong environmental, private sector and developed country interests to establish REDD mechanisms quickly" (Peskett 2008). One particular concern is the rollback of successful decentralised forest management: „With billions of dollars at stake, governments could justify recentralization by portraying themselves as more capable and reliable than local communities at protecting national interest" (Phelps, Webb and Agraval 2010: 312-13). In a similar vein, a study by FERN points to the danger that „rushing REDD processes" will undermine the need for time-intensive governance and consultation processes that are part of other policies, and „will not necessarily address the underlying causes of deforestation" (Leal Riesco and Opoku 2009).

Concerns are particularly strong for market-based schemes that „might suffer from greater efficiency-equity trade-offs" (Peskett 2008), as „paying for a carbon service [...] makes it harder to incorporate issues such as biodiversity and poverty considerations" (Scholz and Schmidt 2008). Proposals for so called market-linked schemes therefore aim at lowering this influence by using intermediary banks or funds, or a separate REDD trading scheme in which Annex I countries commit themselves to purchase REDD credits (Schmidt and Scholz 2008).

5   Investments in climate politics – some alternative framings

Finance has become a crucial issue in current climate politics: The level and direction of finance flows will be decisive for the speed and scope of the transformation to low-carbon economies, and play a crucial role for adaptation and clean development in developing countries. Equally, there can be no doubt that private investments will have a leading part here, as they account for 86 per cent of global financial flows (UNFCCC 2007).

Against this background, it is rather surprising that the role of investment flows in climate politics has not been addressed explicitly at an earlier point in time. The way in which the issue was taken up in recent years, however, results. in a narrow framing that focuses on incentivising new investments, instead of starting from the desired political outcomes and asking for the appropriate regulation to achieve these objectives. This framing can only be understood within the context of a general economic understanding of climate change and climate politics that allows to reduce the complexity of climate politics to a range of abstract abatement opportunities. The following will highlight the most problematic consequences of this understanding and suggest alternative framings.

First, the very call for massively scaling up the level of climate finance, using all available sources of money, is based on a problematic reduction. Estimating that it will cost US $ 380 billion in 2030 to return emissions to a 2007 level, the UNFCCC secretariat emphasises that this sum is only a fraction of 1.1 to 1.7 of total investment and financial flows in 2030. This relation shows that the problem consists less in an absolute scarcity of money but rather in the use of current financial flows, a perspective that is shared within the investment community: „Institutional investors are searching for new asset classes and strategies [...] and the climate economy is emerging as an attractive source of long-term returns" (Robins and Fulton 2009).

Second, one consequence of the run for scaled up climate finance is the non-discrimination of different sources of money. While it may be true in macroeconomic terms that purchasing offset is from the viewpoint of developed countries „just as much an expenditure of societal resources as ODA and other public financing mechanisms" (Stewart et al 2009: paper, 3), the form of financial transfers definitely makes a difference for the recipients: Offsets are subject to the investors' rationality of maximising returns, in this case the number of certificates, while public spending can follow political objectives. And the use of public money is crucial in important areas of climate politics, like the adaptation of the most vulnerable groups and societies (Stern 2009, Gore 2010). The global level of climate finance flows is therefore an insufficient indicator for what is needed. Rather, it is important to ensure that the policies and measures that are most urgently required receive sufficient financial support.

A third crucial logic in the current climate finance debate is the priority setting along cost-effectiveness that results in a higher share of developing countries in global emission reductions. This logic is derived from the same macroeconomic considerations that characterise the Stern Review, and relies on a similar reduction of complexity. A look at the REDD debate shows that this logic takes effect in current climate politics. A forest finance mechanism seems to offer the ideal case of a win-win opportunity: While developing countries hope for large financial inflows, many developed countries seem to expect that markets and private investors will soon step in and lower the burden for national budgets.

These expectations seem kind of paradoxical given the many concerns raised not only by NGOs, and the uncertainties regarding the implementation of REDD, as discussed within the UNFCCC. While this must not be a reason to lower the ambition to a mechanism for reducing deforestation, it questions the assumption that REDD can offer a bridge strategy to postpone emission reductions in particular in developed countries. More realistic expectations to the potential and scale of REDD could offer more fruitful conditions for implementing the mechanism in a way that considers the interests and concerns of all stakeholders and most affected groups, instead of sidelining these issues for financial reasons.

The emphasis on financing emission reductions in the forestry sector, however, implies a focus on the places where deforestation occurs, and on the forestry sector itself, whereas it is widely acknowledged that „most of the underlying causes lie outside the forestry sector" (Scholz and Schmidt 2008). Addressing these causes would require international action beyond the UNFCCC, that „should include revision of international trade policies, introduction of socioecological import standards and reform of agro-fuel quotas and other relevant policies in the EU, the US and elsewhere" (Scholz and Schmidt 2008). Rather than suggesting that „Host countries should identify investment opportunities and partnerships with international investors", ensuring „that the interests of local communities are aligned with REDD+ projects" (Romani et al. 2009: 17), „the approach needs to be turned on its head, and the mechanism subordinated to the problems that it is trying to address" (Brown and Bird 2008).

This leads to a final and central issue, the role of private finance and the need for incentivising investments. Here, a much more differentiated perspective is needed. It is certainly the case that developing countries need financial support not only for adaptation, but for pursuing cleaner development paths that result in lower emissions. And it follows directly from the large share of private money in global financial flows that much of the required investment will have to come from the private sectors.

The current climate finance debate, however, focuses narrowly on generating additional financial resources and supporting investors. Much lesser emphasis is given to the role of existing financial flows. „The problem is often characterized as ... finding a large pot of money quickly to fill the „finance gap"" (Hamilton 2009). Redirecting current and future finance flows does not only offer this large pot of money, but could tackle some of the root causes of climate change at the same time.

That does not mean that instruments like PFM should play no role in incentivising investments for desired outcomes. From a societal perspective, however, it makes a huge difference if public resources are used to support local entrepreneurs, as suggested by some development banks, or to lift the returns for large institutional investors to market levels, as intended by the PFMs proposed in the „meeting the Climate Challenge" report.

Investors call for „regulatory certainty" (Robins and Fulton 2009) and „long, loud and legal" policy signals that minimise political risk (Hamilton 2008): *loud* to make investment more attractive; *long* to reflect the financing horizon of a project; and *legal* to meet investors concerns on changing regulation. The question arises, however, whether these signals inevitably must take the form „investment grade policies" and „incentive frameworks" (Hamilton 2009) designed according to investor interests; or whether more traditional forms like standards and carbon taxes could play a very effective role here.

Investors who are interested in a strong and clear investment perspective for renewable energies and concerned if governments are „really committed to decarbonising the economy" (Mainelli et al. 2009), could gather their allies around the world to make an equally clear and strong plea for ambitious reduction commitments at the next UNFCCC meetings, and for translating these commitments into strong regulation through standards, carbon taxes or an ambitious emissions cap. Making the use of fossils fuels and the respective technologies more expensive would enhance the attractiveness of investments into renewable energies and clean technologies in a foreseeable and reliable way.[17]

Instead of increasing government´s debt burdens through new financial market instruments like indexed bonds, carbon taxes or the auctioning of emis-

---

[17] By and large, a cap and trade system that deserves the former part of its name by setting an ambitious emission cap would have the same effects, if emission rights are auctioned.

sion rights would generate large amounts of public finance that could be used for many of the policies and measures that are beyond the scope of investors, but are very desirable from a societal point of view.

If it is generally accepted that through the environmental and financial crisis a „rather radical redefinition of the role of government in the market place, and the role of public policy in protecting public interests, feed through into a more active approach to government regulation in climate and energy" (Hamilton 2008: 5), and even investors conclude that „Increasingly climate change is being viewed as another example of systemic risk failure on capital markets, with the failure to adequately price carbon being compounded by incentive-driven short-termism" (Robins and Fulton 2009: 146), the question arises whether the active role of governments has to limit itself to providing preferential investment environments, or could contribute to a much broader transformation.

Instead of accepting a „Decarbonised Dystopia" (Newell and Patterson 2010) in which large scale energy investment projects deliver low-carbon energy supply through carbon markets but have no or only marginal egalitarian effects, the large levels of public spending that are required in coming years and decades could be used for addressing some of these inequalities, in the national as much as in the north-south context. One fundamental step in that direction would be, as Lohmann (2009) suggests, to strengthen the role of societies in making investment decisions, and to aim at „locally-focused" energy systems that are oriented in the needs of populations. This alone would not mean to overcome „climate capitalism" (Newell and Patterson 2010), but could contribute to questioning the influence of global financial markets on political decisions rather than further enhancing it.

References

Ackermann, F. (2007): Debating climate economics: the Stern review vs. its critics. Report to Friends of the Earth-UK. Meford: Tufts University.

Andersen, A. (2008): Moving ahead with REDD: Issues, options and implications. Bogota: CIFOR.

Arrow, K. (2007): Global climate change: a challenge to policy. Economists Voice, 4 (3).

Bäckstrand, K.; Lövbrand, E. (2006): Planting trees to mitigate climate change: Contested discourses of ecological modernization, green governmentality and civic environmentalism. Global Environmental Politics 6 (1): pp. 50-75.

Barker, T. (2008): The economics of avoiding dangerous climate change. An editorial essay on The Stern Review. Climatic Change, Vol. 89: pp.173–194.

Bettelheim, E. (2009): Forest and Land Use Programs must be given financial credit in any climate change agreement. In: Stewart, R. et al. (2009): Climate Finance: Key concepts and ways forward: Harvard Project on International Climate Agreements.

Bond, I. et al. (2009): Incentives to sustain forest ecosystem services: A review and lessons for REDD. Natural Resources Issues No.16. London: IIED.

Brinkmann, M. (2009): Incentivizing private investment in climate change mitigation. In: Stewart, R. et al.: Climate Finance. Regulatory and Funding Strategies for Climate Change and Global Development. New York/London: New York University Press, pp. 135-142.

Brown, D.; Bird, N. (2008): The REDD road to Copenhagen: Readiness for what? London: Overseas Development Institute.

Cline, W. (1992): The Economics of global warming. Washington: Institute for International Economics.

Doornbosch, R.; Knight, E. (2008): What Role for public finance in international climate change mitigation. OECD Discussion Paper. Paris: OECD.

Dutschke, M.; Wertz-Kanounnikof, S. (2008): Financing REDD. Linking country needs and financing sources.

Egner, H. (2007): Überraschender Zufall oder gelungene wissenschaftliche Kommunikation: wie kam der Klimawandel in die aktuelle Debatte? *GAIA*, Vol. 16, No. 4: pp. 250–254.

Eliasch, J. (2008): Climate Change. Financing Global Forests. London: Earthscan.

European Commission (2009): White Paper: Adapting to climate change: Towards a European framework for action.

Fankhauser, S. (2006): The Economics of Adaptation. Available at www.hm-treasury.gov.uk/d/stern_review_supporting_technical_material_sam_fankhauser_23100 6.pdf, accessed May, 20th 2009.

Foucault, M. [1976] (1998). The History of Sexuality Vol. 1: The Will to Knowledge. London: Penguin.

Gomez-Echeverri, L.; Müller, B. (2009): The financial mechanism of the UNFCCC. A brief history. Oxford: European capacity building initiative.

Gore, T. (2010): Climate Finance Post Kopenhagen. Brussels: Oxfam.

Hamilton, K. (2008): Clean Energy Investment and the „New Competitiveness". London: Chatham House.

Hamilton, K. (2009): Unlocking Finance for Clean Energy: The Need for „Investment Grade" Policy. London: Chatham House.

Iola Leal, R.; Opoku, K. (2009): Is REDD undermining FLEGT? Avoiding Deforestation and Degradation Briefing Note 5, March 2009, FERN.

Jessop, B. (1999): The Governance of Complexity and the Complexity of Governance: Preliminary Remarks on some Problems and Limits of Economic Guidance. Lancaster University. Avaliable at http://www.lancs.ac.uk/fass/sociology/papers/jessop-governance-of-complexity.pdf.

Karousakis, K.; Corfee-Morlot, J. (2007): Financing mechanisms to reduce emissions from deforestation: issues in design and implementation. Paris: OECD and IEA.

Lubowski, R. (2008): The role of REDD in stabilising greenhouse gas concentrations. Lessons from economic models. Bogota: CIFOR.

Levy, D.; Egan, D. (2003): A Neo-Gramscian Approach to Corporate Political Strategy Conflict and Accommodation in the Climate Change Negotiations. Journal of Management Studies 40, no. 4: pp. 803-29.

Lohmann, L. (2009): Climate as investment. Development and Change 40 (6): pp. 1063–1083.

Luks, F. (2008): Der Diskurs über das Klima und das Klima des Diskurses. GAIA, Vol. 17, No. 2: pp. 186–188.

MacKenzie, D.; Millo, Y. (2003): Constructing a Market, Performing Theory: The Historical Sociology of a Financial Derivatives Exchange. American Journal of Sociology, 109 (1): pp. 107–145.

Maclean, J. et al. (2008): Public Finance Mechanisms to mobilize investment into climate change. Paris: UNEP.

Martone, F. (2010): The emergence of the REDD hydra. An analysis of the REDD-related discussions and developments in the June session of the UNFCCC and beyond. Forest Peoples Programme.

Mc Kinsey (2009): Pathways to a low carbon Economy. London.

Newell, P.; Patterson, M. (2010): Climate Capitalism. Cambridge: Cambridge University Press.

Okereke, C.; Bulkeley, H.; Schroeder, H. (2009): Conceptualizing Climate Governance Beyond the International Regime. Global Environmental Politics 9, no. 1: pp. 58-78.

Parker, C. et al. (2009): The little Climate Finance book. London: ODI/GCP.

Pendleton, A.; Retallack, S. (2009): Fairness in Global Climate Change Finance. London: Institute for Public Policy Research (IPPR).

Peskett, L. (2008): Making REDD work for the poor. London: ODI.

Phelps, J.; Webb, E.; Agrawal, A. (2010): Does REDD+ Threaten to Recentralize Forest Governance? Science, Vol. 328: pp. 312-313.

Project Catalyst (2009): Scaling up climate finance. Finance briefing paper. Available at http://www.project-catalyst.info/images/publications/climate_finance.pdf, accessed February, 28th 2011.

Price, C. (1993): Time, Discounting and Value. Oxford: Basil Blackwell.

Raworth, K. (2007): Adaptation to Climate Change. What´s needed in poor countries and who should pay. Oxford: Oxfam International.

Robins, N.; Fulton, M. (2009): Investment Opportunities and Catalysts. Analysis and proposals from the Climate Finance Industry on Funding Climate Mitigation. In: Stewart, R. et al.: Climate Finance. Regulatory and Funding Strategies for Climate Change and Global Development. New York/London: New York University Press, pp. 143-151.

Romani, M. et al. (2009): Meeting the Climate Challenge: Using Public Funds to Leverage Private Investment in Developing Countries. Summary for policy makers. London: London School of Economics.

Rubio Alvarado, L.; Wertz-Kanounnikof, S. (2007): Why are we seeing REDD? An analysis of the international debate on reducing emissions from deforestation and degradation in developing countries. Paris: IDDRI.

Schalatek, L. (2009): Gender and Climate Finance. Washington: Heinrich Böll Foundation.

Schneider, S. (2008): The Stern Review debate: an editorial essay. Climatic Change, Vol. 89: pp. 241–244.

Scholz, I.; Schmidt, L. (2009): Financing the climate agenda: the development perspective. Background Paper to "International policy dialogue – Financing the climate Agenda". Berlin, March, 19th – 20th 2009. Bonn: DIE-GDI/BMZ.

Spash, C. (2007): The economics of climate change impacts à la Stern: Novel and nuanced or rhetorically restricted? Ecological Economics 63: pp. 706-713.

Stern, N. (2006): The Economics of Climate Change. The Stern Review. New York: Cambridge University Press.

Stern, N. (2009): Blueprint for a Safer Planet: How to Manage Climate Change and Create a New Era of Progress and Prosperity. Random House.

Streck, C. (2009): Expectations and Reality of the Clean Development Mechanism: A Climate Finance Instrument between Accusation and Aspirations. In: Stewart, R. et al. (2009): Climate Finance: Key concepts and ways forward: Harvard Project on International Climate Agreements.

Stewart, R.; Kingsbury, B. (2009): Climate Finance: Key concepts and ways forward. Harvard Project on International Climate Agreements.

The Climate Group (2009): Cutting the cost. The economic benefits of collaborative climate action. Cambridge University.

UNFCCC (2007): Investment and financial flows to address climate change. Bonn.

Weitzmann, M. (2007): A review of The Stern Review on the economics of climate change. Journal of Economic Literature, Vol. 45: pp.703–724.

Wolf, S. (2009): Reducing problems through reduced complexity? Considering the benefits and limits of economic perspectives on climate change. Int. J. Green Economics, Vol. 3, Nos. 3-4: 304–322.

World Bank (2006): Managing climate risk. Integrating adaptation into World Bank operations. Washington D.C.: World Bank.

World Bank (2008): Proposal for a strategic climate fund. Design Meeting on Climate Investment Funds. CIF/DM.2/3. Washington D.C.: World Bank.

World Bank (2009): The Global Report of the Economics of Adaptation to Climate Change Study. Washington D.C.: World Bank.

Yohe, G.; Tol, R. (2008): The Stern Review and the economics of climate change: an editorial essay. Climatic Change, Vol. 89: pp. 231–240.

Contradictions of the Commodity Carbon – On the Material and Symbolic Production of a Market

Martin Bitter

1 Introduction

It can be attested that the capitalist mode of production has at least two breath-taking, irresistible potencies. The first expresses itself in the dynamics, in the expansive character of the capital form. Processing values continually return enhanced to themselves and generate growth which is quantifiable in monetary terms. Social labour power continually achieves new points of culmination through technological and organisational innovations, and natural material is transformed into ever new use values for the satisfaction of human needs. The second, tendentially repressive force begins to take effect as soon as the process of accumulation begins to falter or even comes to a halt. This is the force of restructuring through crisis. The producers take their privately appropriated commodities to market but the commodities cannot be sold there. There is no solvent demand. Payment chains break and credits can no longer be repaid. The downward spiral begins to turn, with all its economic, social and political consequences.

Capitalism is thus creative and destructive in equal measure, and it is so not only in relation to the societal (class) relationships on which it is based but also in it relationship to nature. One of the first to attempt to synthesise these two aspects and to translate them into an ecological theory of the capitalist mode of production was the US-American economist and sociologist James O'Connor. In his paper „Capitalism, Nature, Socialism" published in 1988 O'Connor describes two types of crisis that are inherent in the capitalist mode of production. The first, which he labels the „first contradiction of capitalism", arises from the coming to a head of the contradiction between productive forces and relationships of production. It results in periodically recurring „crises of overproduction". The principle of the expropriation of surplus value turns upon itself. Commodities cannot be converted into money through exchange and the rate of profit falls. The „second contradiction of capitalism", on which an „ecological Marxism" could be based, aims its sight at the relationship of tension among productive forces, relationships of production and production conditions. The latter describe those requirements of capitalist production which make the social process of reproduction possible in the first place. As a rule, however, these cannot be provided using individual capitalist strategies of valorisation. Among them O'Connor counts labour power, natural preconditions (fertile land, mineral resources, fossil energy sources etc.) and the general conditions of production, i.e. in particular transport and communication infrastructures. As far as the conditions of capitalist production are concerned, capital has an inherent tendency to continually undermine them. It treats them as if they were unlimitedly available, as if they could be consumed without having to be previously produced. O'Connor argues that the conditions of capitalist production are no longer available in the necessary quantity and quality, and he therefore speaks of an „underproduction"[1] of the conditions of production.

Since autumn 2008 global capitalism has found itself in a situation in which the contradictions elaborated by O'Connor overlap. The contradictions between price and wage deflation on the one hand, and interest rate and yield inflation on the other (Altvater 2006: 122, Harvey 2010: 26ff.) have finally given vent to a crisis in the capitalist centres. The unwavering, continued burning of the fossil ‚lubricant' of capitalist development has, in contrast, made clear the urgency of political action on the question of anthropogenic climate change. The latter cannot be adequately grasped, however, without the „first contradiction" – that is the basic thesis of this paper. The decisive question is how over- and underproduction are connected, how they *articulate* themselves within the framework of the field of immanence of the capitalist mode of production. O'Connor's dualist perspective (cf. Castree 2002: 124f.) must be transformed in order to answer this. It must be transformed in such a way that nature and society are understood as a contradictory unity that takes on particular forms of movement in specific historical contexts.

The theorem of „ecological modernisation" (Mol 2003) is paradigmatic for the present, (post-)Fordist phase of the appropriation of nature by society (Brand/Görg 2003). It describes the attempt to adjust the capitalist accumulation imperative – conveyed by an efficient state market design – such that the pollution of the environment can be reduced to an „optimal amount". The central

---

[1] O'Connors concept of „underproduction" transports an idea, however, which he actually rejects: the reproducibility of natural conditions of production (cf. Altvater 1992: 285ff.). Underproduction is, after all, only possible where production takes place. The use of fossil energy sources, however, describes a process which from a human perspective is irreversible (if one abstracts from the geological ‚production process' which lasted millions of years). A barrel of oil can only be used once. Underproduction, in contrast, suggests the circularity and reversibility of processes which by their nature can only take place once.

vehicle of this strategy is the pricing or valorisation of the exploitation of nature, which is generally referred to as the „internalisation of external effects". In contrast to this, the current paper aims to unearth the contradictory character of this new generation of commodities from an immanence perspective. It illustrates this using the example of the European Union Emissions Trade System (EU ETS). In a first step the theoretical framework of the „production of nature" (Smith 1996) will be elaborated. Following that, the strategy of ecological valorisation will be explained and its contradictory character will be examined with reference to emissions trade.

2 The material and symbolic production of nature

The idea of understanding the relationship between society and nature from an immanent perspective describes first of all a quite simple ontological fact: man, himself a natural being, continually affects as a sensory creator the nature which is external to him. In order to survive he directs his skills toward transforming nature and adapting it to his own needs. Wherever man affects nature there can therefore be no untouched, pristine conditions. At the same time, however, man not only appropriates nature and makes it subject to himself by making its laws work for him. Since he is himself a natural being, and his sensory activity is after all only possible because of this, nature in its intractability affects man in return. It changes him, pre-forms his path of development and sets both relative and absolute limits.

This relationship of tension – Marx' argument can be used here – can be understood specifically using the example of the category of *labour*. Labour represents the unshakeable, overhistoric human necessity of the appropriation of nature for the purpose of the production of use values and the satisfaction of needs: „We shall, therefore, in the first place, have to consider the labour-process independently of the particular form it assumes under given social conditions." (Marx, Capital, Vol. I: 177) Beyond that, however, the labour process is always also embedded in a particular societal mode of production. In the capitalist mode of production the labour process takes on the specific social form of wage labour and thus assumes a „two-fold character" (ibid.: 41). Labour is *firstly* the creator of value. In it is expressed the abstract norm of socially necessary labour time. Value is therefore a social relationship which contains „not an atom of matter" (ibid.: 47) but which has a structuring effect on the activities of the actors. *Secondly*, it is also „a special productive activity, exercised with a definite aim, an activity that appropriates particular nature-given materials to particular human wants" (ibid.: 42). Abstract value-creating and concrete useful labour are there-

fore two aspects of one and the same labour process. For the articulation of natural and societal factors it is therefore not only important that a transformation of
matter and energy for the purpose of the production of use values takes place. It
is decisive that this happens under the „historical ‚template' of the law of value
and the procedures of the market" (Altvater 1992: 250). Since in the capitalist
mode of production things are produced for the purpose of exchange, we must
abstract from the fact that the product of labour assumes its form through concrete activities aimed at changing its quality. The „spherical dependence" (ibid.:
242) of production and consumption, their use value aspect goes under in the
wake of the self-referential socialisation via the exchange value.

This does not remain without consequence for the societal forms of the appropriation of nature, however. Since the capital form assumes an „automatically
active character" (Marx, Capital, Vol. I: 153) and constantly attempts to create
new wants in ever shorter periods of time, the circulation between man and nature becomes successively dynamic: „Today, the ‚second nature' in a Hegelian or
Marxist sense is increasingly less often produced from (and in contrast to) ‚first
nature'; rather, the ‚first nature' is produced from, and as part of, the second
nature." (Smith 2008: 876) The processual character, the continual transformation and upheaval of socio-natural constellations become decisive. On the one
hand, this can be shown quantitatively, i.e. with reference to the rate of growth of
gross domestic product. The relationship of capital, fuelled by fossil energy and
coupled with the concept of a „life-style devoted to one's calling" (Weber 2006:
78) based on the idea of vocational duty led between 1820 and 1998 to the ten-
fold multiplying of the rate of growth of average annual income per capita from
0.22% to 2.21% (Maddison 2001). The historically singular climax of this development was the Fordist „gospel of consumption" (Hunnicut, quoted in Panayotakis 2006: 268). Between 1950 and 1973 the rate of growth was just under
5%.[2] On the other hand, resulting from the two-fold character of the labour process the valorisation process also has qualitative aspects. Thus, the accumulation
process was always accompanied historically by an increased throughput of
matter and material, which led among other things ecologically to the (from the
perspective of valorisation) irreversible entry of greenhouse gases into the atmosphere. In addition, the accelerated, socially formed re-formation of natural
materials to use values has given rise to new social relationships of work and

---

[2] Of course, from a socio-ecological perspective it is decisive to take into account that the absolute
growth of economic output, which for a constant relative rate of growth must be continually higher.
Maddison estimated the total output of goods and services of the world economy in 1820 at $694
billion (at constant dollar values of 1990), in 1950 the figure was $5.3 trillion, in 1973 $16 trillion, in
2003 $41 trillion and in 2009 – according to the World Bank Development Report – $56.2 trillion
(Maddison 2001; Harvey 2010: 26f.).

living with a fundamentally changed spatiotemporal matrix. Socio-natural configurations have arisen which structure social relationships. Patterns of urbanisation as well as transport and communication infrastructures represent „created ecosystems" (Harvey 1996: 186) or „nature-culture-hybrids" (Latour 1993), which characterise the subsequent development options of the material production of nature and give them structure. Last but not least, this has consequences for the mode of operation of the capitalist state. Once it has been ‚enticed into' the Fordist development paradigm, the fossil-driven growth fetish becomes a fundamental characteristic of each and every power strategy. From this point on, the supporters of a societal mode of development removed from growth are representatives of an irrational or immoral knowledge (cf. Jessop 1990: 208). The „strategic selectivity" (Jessop 2008: 36)[3] of state institutions therefore results in priority being given in particular to the business areas of the energy sector. Because of the close relationship between the use of (fossil) energy and economic growth, this sector can lay claim to being the representative of general capital interests (Newell/Paterson 1998).

Put in a generalisable formula: forms of the societal appropriation of nature or the material production of nature contain not only the „totalitarian moment" (Swyngedouw 2009: 386) of an irreversible socio-ecological intervention. In the form of new „created ecosystems" they also revolutionise a society's forms of political organisation. They transform the forms of work, of political participation, of distribution, of value orientations with all their spatiotemporal coordinates. „All ecological projects (and arguments) are simultaneously political-economic projects (and arguments) and vice versa. Ecological arguments are never socially neutral any more than socio-political arguments are ecologically neutral. Looking more closely at the way ecology and politics interrelate, then becomes imperative if we are to get a better handle on how to approach environmental/ecological questions." (Harvey 1996: 182) In the formation of an immanent, relational concept the ecological dimension of politico-economic developments must therefore be taken into account, while at the same time this leaves no room for explicit ‚environmental policy'. The latter cannot be separated from its implications for the effects of social distribution, the democratic content and the politico-economic relationships of power and dominance.

This does not imply, however, that societies cannot be capable of critically questioning the forms of the appropriation of nature in social discourses and of

---

[3] Jessop (2008: 36) understands the „strategic selectivity" of state institutions as follows: „It (the state system, M.B.) can be analysed as a system of *strategic selectivity*, i.e., as a system whose structures and *modus operandi* are more open to some types of political strategy than others. Thus a given type of state (…) will be more accessible to some forces than others according to the strategies they adopt to gain state power."

naming them for what they are. Societies also produce nature symbolically and linguistically. They set out on contested, contingent processes of searching in order to satisfy themselves with regard to their relationship to external nature. Nature, however, does not per se provide a positive criterion, an objective norm, which could place a framework of ‚prestabilised harmony‘ on societal relationships with nature (Morton 2007). Possibly undesirable developments and the limits to the appropriation of nature must therefore be identified by means of scientific, cultural, economic and political interpretations. In the arenas of the political state and civil society – i.e. in the framework of that which Gramsci described as the „integral state“ (Gramsci 1991ff., No. 12, § 1: 1502) – their justification can be probed and, if found suitable, cast as hegemonial patterns of the appropriation of nature into the form of political commitment. The patterns cannot be chosen arbitrarily, however. Components of the symbolic-linguistic construction are always also translated back into the material production of nature. This can then, last but not least, demonstrate to man the failure of social strategies of the appropriation of nature (Görg 2003: 119f.). Anthropogenic climate change is only one example of the fact that societies are urged by intractable nature to a „second reflection“ (Adorno 1982: 186, quoted in ibid.: 123) of their linguistic-symbolic production of nature.

The symbolic-linguistic production of nature must be differentiated from the social-constructivist view of nature, „the concentration of which on discourses replaces a sound analysis of social production or of political-social economy instead of complementing it“ (Smith 2008: 875). The symbolic-linguistic and the material production of nature cannot be separated from one another as they mutually constitute each other and cannot be dissolved in the direction of one of the poles. The „antithesis constructivism-realism“ can therefore „only be overcome if both poles are radicalised at the same time“ (Görg 2003: 121). We are dealing with a contradictory internal relationship which evolves dialectically from a unity. The analytical separation of the material and linguistic-symbolic production of nature thus presents itself, in a manner of speaking, as the foundation and the superstructure of societal relationships with nature, knowing full well that it – just like Marx’ original model – is not amenable to a deterministic interpretation or to linear causality.

3   Valorisation and „occidental rationalism“

What exactly does the „second reflection“ on the social appropriation of nature, which has been triggered by anthropogenic climate change, look like? What concept of nature is linguistically and symbolically produced here? To what extent is the immanent relationship, the co-evolution of society and nature, taken into account?

Let us begin with the fundamental cognitive paradigm, with the intellectual capacity which western societies bring into position in order to react to disturbances in their interaction with nature. Max Weber's concept of „occidental rationalism" proves helpful here. This examines the question as to why the forms of life-style in the capitalist centres were canalised into the trajectory of a means-and-ends calculation. In this connection Weber places the focus of his analysis on the re-organisation of the temporal structure of societies. He argues that the moments of daily life are segmented and then re-assembled in such a way that they are condensed into ‚elements of profit'.[4] Acceleration becomes *movens*, embodied in the „capital calculation" (Weber 1980: 48), in the rationalised form of the deployment of the factors of production. Calculability and profitability are therefore also key quantities in societal processes of self-reassurance. They play a major role in the question of how the treatment of external nature can be renewed. From the point of view of „occidental rationalism" this means the improved ability to understand the prognostication and governability of the intractable external. It therefore develops a series of sophisticated models which make climate change approachable from its own viewpoint. Beginning with the founding of climate change research institutes, through the linking of international research networks with the governmental level, to the establishment of policy networks oriented towards the solution of problems – „organic intellectuals" (Gramsci 1991, No. 12, § 1: 1497)[5] in the „integral state" develop strategies for the re-formulation of societal relationships with nature. These discourses, however, also transport the (fetishised) limits of the means-and-ends calculation since, as a rule, they are aimed at dealing with the „greatest failure of the market in the history of mankind" (Stern 2007), anthropogenic climate change, with the traditional social mechanisms of discovery. Nature is given a price and becomes a commodity. Ecological problems are to be „internalised" in the economic calculation and function as a cost factor to sink the entropic production rate[6] of capitalist societies, or rather: to make them more efficient.

---

[4] Cf. the fundamental importance which Weber attaches to Benjamin Franklin's dictum: „Remember that *time* is *money*" (Weber 2006: 75). Weber argues here in implicit agreement with Marx' reflection that the value of a commodity is measured by the amount of socially necessary labour *time* it contains.

[5] Gramsci depicts as „organic intellectuals" the producers of ideologies whose concepts, projects and arguments lend the existence of the bourgeoisie a particular, sensory outline and characterise the latter's self-image, while they consolidate or transform the collective life-style of their subordinates.

[6] According to the second law of thermodynamics, entropy (the dispersal of energy) is increased in an isolated system by the process of the transformation of matter and energy, resulting in an irreversible decrease in the quality and thus the usability of energy.

As a dominant political project the concept of coming to grips with so-called environmental problems[7] by means of economic cost-benefit calculations and the functioning of market mechanisms is quite new. It was not until the end of the 1980s that those (optimistic) voices could no longer be overheard which claimed that environmental problems could be dealt with adequately without having to encroach upon the foundations of western norms of production and consumption along with their attractive life-style. „New politics of pollution" (Weale 1992) and „ecological modernisation" (Mol 2003) were the buzz-words for the assumption that economic growth (the increase in the quantity of com-modities per unit of *time*) and the protection of the environment could in fact be reconciled. The position of the Club of Rome on the „Limits to Growth" (1972), which the environmental movement of the 1970s had made its own, was coun-tered with the argument that the market could be regulated in such a way that in the end something like the „optimal" rate of environmental pollution would be the result. A key document for this position was supplied by the British econo-mist David Pearce with his book „Blueprint for a Green Economy" (1989) – the so-called Pierce Report. Pierce proposed translating the concept of „sustainabil-ity"[8] which had been expounded in the „Brundtland Report" (1987) into econom-ic activity via a strategy of ecological valorisation.

Pierce's strategy has in the meantime been quite successful. Man's external nature, however variegated and diverse its individual manifestations may be, has already often been reduced to a qualitatively uniform measure, namely to a mon-etary one. It has been transformed into a tradable commodity without, however, forfeiting its biophysical singularity. Trade takes place with CO_2 certificates and with biodiversity (McAfee 1999), just as with water (Bakker 2005), fishing quo-tas (Mansfield 2004) and wetlands-certificates (Robertson 2004). And why not? Was not one of the central problems of the over-exploitation of nature that nature was available free to the economic actors? No price had to be paid for it. Its sustainable use could be ignored with a smile. Was not the essence of the „trage-dy of the commons" (Hardin 1968) that the benefit-maximising individual pre-cisely did not serve – as with an invisible hand – the ‚welfare of all', but instead produced a ‚public bad' (environmental damage and the exhaustion of re-sources)? The Stern Review (2007) begins with this point. Finely engraved in the concepts of cost-benefit analysis Stern follows Weber's dictum that in the „ent-mystified world" in principle one „could *master* all things by *calculation*" (We-

---

[7] Concepts such as „environmental policy" and „environmental problems" transport the assumption of the principle externality of society and nature, with the result that their mutual constitutional relationship tends to disappear from focus (cf. Görg 2003: 121f.).
[8] Sustainability is defined in the Brundtland Report as „development that meets the needs of the present without compromising the ability of future generations to meet their own needs".

ber 1973: 317). He argues that even a relatively small volume of investment in the „internalisation of external effects" (1% of GDP) would mean that the long-term economic damage of climate change could be kept within limits and the path of economic growth perpetuated.[9] Stern speaks the rational language of money, a language which avoids moral judgement of socio-natural processes and at the same time touches the everyday understanding of each and every individual. Everybody possesses and uses money, and everybody has a practical and intellectual concept of what money actually is (Harvey 1996: 150f.). Money represents the only *universal* measure of value in the capitalist formation of society. It is a ‚radical equaliser'. By means of valorisation, different encroachments on nature (the construction of a dam or the maintenance of the biodiversity of a rainforest) can be made comparable and their monetary justification can be probed. The „Berufsmensch" can in this way grasp the „tragedy" he or she has caused as part of the „capital calculation". This is the strength of the language of money: the socially dominant groups, as those ‚mainly responsible' for environmental damage, and the great majority of society with their monetarily schooled everyday understanding can comprehend it straightaway. In the course of this, the „Berufsmensch", whose life is oriented towards his calling, perceives that the processes of valorisation and market creation are turning nature into an interesting object of investment. It represents a rational investment decision, an accumulation strategy (Smith 2006; Lohmann 2009; Bumpus/Liverman 2008).

3.1 Power and private ownership

Before qualitatively different components of man's external nature can be treated as exchangeable equivalents, i.e. given a price and offered for sale on the market, however, certain conditions have to be fulfilled. The new ecological commodities must first be constructed by means of state intervention (Peck/Tickel 2002; Castree 2008). This means, firstly, that private property rights or rights of ownership over nature must be awarded. After all, the central idea behind monetarisation is to limit the overexploitation of nature by creating an artificial scarcity of the resources demanded. Environmental damage is reduced to an ‚allowed' amount sanctioned by the state. It therefore requires exclusive private rights. It has to be defined who has access to nature under what conditions and for how long, and who does not. The privatisation of nature thus influences not only the way in which humans appropriate nature. It also constitutes a social relationship

---

[9] Cf. also the McKinsey Report „Pathways for a low-carbon economy" (2009), which estimates the costs to the economy as a whole of successful strategies of emission avoidance and adjustment to climate change as being much lower than the estimates of the Stern Report.

between owners and those excluded from ownership. Privatisation generates relationships of power (Lohmann 2006: 78; Altvater 1992: 278). It is of considerable importance which part of man's external nature is to be valorised. It would, for example, be highly impractical to award private property rights for carbon dioxide. CO_2 has no use value. Moreover, it is poisonous. It does not satisfy any need. Nobody would buy it. The market could not function as a social finding mechanism. This is why certificates are awarded which permit the emission of CO_2 („allowances"). Permission is of course only granted for a limited period (in the EU ETS for one trading period), because the „internalisation mechanism" can only be successful if the artificial scarcity is intensified step by step. The CO_2 certificates thus in a manner of speaking represent „semi-permanent property rights" (Lohmann 2006: 81).

If we now concentrate our attention more closely on the negotiation and implementation process of the creation of a market in the multilevel system of the European Union, we can recognise two analytically separate logics of valorisation (Bailey/Maresh 2009). One of these is the regulatory logic, i.e. the hegemonial interpretation of emissions trade. This is based on the „hope of a virtuous fusion of economic growth, efficiency, and environmental conservation" (Bakker 2005: 543). With a political project aimed at supporting the EU's claim to global leadership in the arena of international climate policy, the world is to be shown that the reduction of greenhouse gas emissions need not be expensive. On the contrary, it could even generate a number of political and economic advantages. The supranational ‚level playing-field' would only strengthen this trend, as the costs at this level would be significantly lower than at the level of the national state (Oberthür/Tänzler 2007).

The basic consensus achieved in this way with regard to the choice of instruments, however, transports a fundamental contradiction. An ambitious *cap*, i.e. the restrictive handling of the awarding of emission allowances would necessarily fundamentally question the conditions of production in those sectors formed by the central industries of the Fordist growth model. The large energy producers, the iron, steel, aluminium, cement and chemical industries[10] therefore endeavoured to present their business interests as those of capital as a whole and to form alliances with ‚their' national governments. The territorial logic of emission trade began to take effect, with the result that the decision as to the total number of certificates (the cap) and the form of their distribution was conceded

---

[10] The aluminium sector succeeded in being excluded from the EU ETS – with the argument that it was disadvantaged by the global aluminium price ascertained by the *London Metal Exchange* compared to non-European competitors (Bailey/Maresh 2009: 451). The chemical industry, represented by the *European Industrial Federation CEFIC*, also succeeded in its plea for exclusion from emission trading (Skjaerseth/Wettestad 2008: 157).

primarily to the Member States (van Asselt 2010: 128). The subsequent overal-location of emission allowances and their free apportionment on the basis of historical emissions („grandfathering") led on the one hand to the collapse of the price of carbon in the first trading period (2005-2007), so that at the end of 2007 the price for one tonne of CO_2 was €0.02. On the other hand the energy producers in particular were able to pocket „windfall profits" by integrating the price of the certificates which they had received for free into their corporate value and by passing on these opportunity costs to the end-user. Solely in Germany these windfall profits are estimated at €6-8 billion in the first trading period (Gilbert-son/Reyes 2009: 36). In this way considerable market power for the most emis-sion-intensive industries was established. RWE, the enterprise with the largest number of allocated emission allowances, received 6% of the total number of certificates. The ten businesses with the largest allocation were able to book one third of the certificates for themselves (Ellerman et al. 2010: 127).

In other words, at least in the first trading period the fossilistic accumulation strategies of the ‚classical' Fordist industries were consolidated rather than chal-lenged. The proponents of emissions trading generally state in their defence that in the first trading period the primary aim was „to get the scheme up and run-ning" (Convery/Redmond 2007: 94). Apart from that, the traditional „strategic selectivity" of Fordist statehood appears to be intact. Now that it has been inte-grated into the institutional framework of the EU ETS, it remains of great im-portance for the latter's future power structure.

The dysfunctionality of the market design arising from the decentralised structure of the EU ETS was a reminder, however, of the urgency of the question as to what would happen if the credibility of emission trading continued to erode beyond the first trading period. What would be the consequence if the EU ETS was not nearly able to fulfil the promises of the economic textbooks? For an EU which laid claim to the leading role in world climate negotiations the answer could only lie in stricter and more cost-intensive direct government instruments. There would be a massive loss of credibility particularly vis-à-vis China and the USA. The rejection of a number of National Allocation Plans by the EU Com-mission at the beginning of the second trading period (2008-2012) and the stronger supranational centralisation of the emission trade for the post-Kyoto phase (2013-2020) thus represent attempts to save the central policy field of EU climate protection policy, namely the carbon market.

The Commission can be sure of the support of the financial industry, for which emission trading has opened up a new field of business. The financial sector can either act as a *broker* and handle the transactions of direct EU ETS participants or it can act as a *trader* and speculate on differences between buying and selling prices. In the last decade a number of firms have therefore been set up whose focus is on trade in CO_2 certificates: Ecosecurities (1997), CO2e.com (2000), Point Carbon (2000). Lobby organisations such as the International Emissions Trading Association (IETA) and the Carbon Markets and Investors Association join forces in order to achieve better political conditions. Major banks such as Barclays and Deutsche Bank are present in the market as are European and US American hedge funds. The latter have held shares in the European carbon market since 2006 (Patterson/Newell 2010; Convery/Redmond 2007: 97f.). These actors are interested in a market design which allows clear, reliable price signals and which is conducive to an increased liquidity of the market. They thus become advocates of an ambitious cap and the auctioning of emission rights. Otherwise the profitability of their involvement would simply be too low and transactions would carry too many risks. Other markets for investments would then be preferred.

The political construction of the EU ETS thus exposes a hybrid regulative framework. On the one hand many aspects of the Fordist era continue to be effective: the major role of fossil energy sources for economic growth, the employment effects of investments in ‚classical‘ industrial segments, the traditional strategic selectivity of state institutions. On the other hand, the „second reflection“ of the societal appropriation of nature has led to the fact that, last but not least, the supranational level of European statehood ascribes priority to the effective political handling of climate change. It is supported by a financial sector whose „structural power“ (Strange 1994) is already recognisable by the fact that the financial markets were granted a fundamental role in the search for cost-effective solutions to the climate problem, an aspect which would scarcely have been imaginable twenty years ago (Newell/Paterson 2009).

3.2 Exchange on markets

The conditions and actors for the second element of valorisation have thus been named – trade and exchangeability on markets (Castree 2003: 279f.). Of course, in the EU ETS no ‚real‘ CO_2 is sold; the material properties of greenhouse gases do not allow this. What is sold is the certificated form of a CO_2 allowance, „paper Co_2“ (Altvater 2006). It is the function of the trade to ensure that emission avoidance is put into effect in the places where it is most cost-efficient. The operators of plants for which a reduction in the emission of greenhouse gases would mean relatively large monetary outlays can buy additional certificates on the market. Market actors for whom the emissions reductions represent „low hanging fruits“ can throw their unneeded certificates onto the market. In order for this exchange to take place smoothly and with the greatest possible informational transparency, a number of financial market actors or „intermediaries" (banks, insurance companies and institutional investors) are engaged between the buying and the selling plant operators. These are intended to guarantee „efficient resource allocation" and to create liquidity using „innovative instruments" such as „strip and swap" deals between EU certificates (EUAs) and other Kyoto certificates (Convery/Redmond 2007). In this process of continually repeated sales of emissions rights and the increase in their market value and volume, a relative decoupling takes place of the material side (‚real CO_2') from the exchange value side (‚paper CO_2') of the greenhouse gases. Certificates circulate on the market without the natural, material side of the trade being immediately recognisable (cf. Smith 2006: 29). The decisive motive in the calculation of the actors is the monetary realisation of the ‚paper CO_2', not the avoiding of greenhouse gas emissions. The price for the emission allowance is ascertained on a politically construed financial market. It is not the result of scientifically proven changes in the chemical composition of the Earth's atmosphere. The decisive parameters are the politically negotiated cap, economic trends, energy prices and the weather (Point Carbon 2004; European Commission 2008). The volatility of the price trend corresponds to the relative independence of the exchange value side (cf. Harvey 1996: 152f.). The monetary form also contains – depending on the specific market design and the behaviour of the actors – the possibility of „market failure". In contrast to other commodities the ‚value' of CO_2 cannot be deciphered as a representative of socially necessary labour time. The monetary evaluation of CO_2 is therefore inherently arbitrary and requires comprehensive state intervention and regulation for its stabilisation, protection and predictability.

So much about the definition of CO_2 trade from the point of view of structural theory. If we look at what happens on the EU ETS market in more detail, it is striking that in particular the interests of financial capital show a contradictory pattern. On the one hand it is strategically and conceptually important to its representatives to allow a market which can show enormous growth rates to continue to mature.[11] They are aware of the possibility that the European engine of a market could develop here, the volume of which „could be comparable with that of credit derivatives within a decade" (Kanter 2007). In day-to-day business, however, it is dominated by the necessity of first providing the deregulated financial markets with a regulative framework on the basis of which profitable trading can be conducted. For example, the IETA is striving to function as a multiplier of scientific expertise on climate change, to popularise the idea of a

---

[11] In the crisis year 2009 the value of the traded European Union Allowances (EUAs) grew due to the exploding volume of trade by 18% to €89 billion (Capoor/Ambrosi 2010). The volume of trade in carbon is still relatively small, however: on the global foreign exchange market daily turnover is $3200 billion (Süddeutsche Zeitung, 29 June 2010: 22).

strict cap and to press for adequate transparency in the auction regulations for the third trading period. Furthermore, the political legitimacy of the financial sector has been damaged since the global economic crisis. The opportunity ‚to wash itself free of its sins' in association with environmental NGOs as the ‚green capital faction' in a social ecological modernisation bloc therefore opens up a perspective for the conservation of its own structural power (see Newell/Paterson in this volume).

On the other hand there is the problem that the „first contradiction of capitalism" has not at all been adequately dealt with by the landslide falls in value of autumn 2008. With a view to the global carbon market David Harvey therefore warns: „Capitalism never solves its crisis tendencies. It simply moves them around. It moves them around geographically, and it moves them around from one kind of difficulty to another."[12] This leads to the question whether the „capital surplus absorption problem" (Harvey 2010: 45) in the global debt economy can be solved in such a way that it can be fixated spatiotemporally in the carbon market. The character of the commodity carbon is without doubt suitable for this: „Unlike other commodities such as oil or grain, they (emission allowances, M.B.) incur neither storage costs nor transport costs. Like financial assets, allowances are mere bookkeeping entries that can be electronically transferred instantly and at very low costs." (Ellerman et al. 2010: 146) The commodity carbon is congruent with the time-regime of the global financial markets, which is based on the continual compression of the turn-around times of capital. Nor do the forms of the trade do anything to counter this acceleration pressure. In the EU ETS in crisis periods around half of all transactions take place *over-the-counter* (OTC), i.e. without the interposition of a trading platform (Capoor/Ambrosi 2010: 20). This guarantees anonymity and allows the breaking down of the transactions into small packages. In this way the exertion of pressure on the price trend can be avoided (Walhain 2007). This increases the opportunities for „gaming in the carbon market" (FoE 2010). In this way energy producers with their market power behind them can buy a large number of EUAs as long as prices are low. They then switch to energy sources with high carbon emissions (e.g. lignite), and subsequently inform the market of the increased emissions profile. Another example is the so-called carousel fraud for avoiding the value added tax on the transactions. This is so easy particularly because no physical goods have to be transported. Option transactions are also strongly on the increase (a growth of 70% in 2009). These are asymmetric forward transactions which speculate on a rising or falling carbon price. And finally the derivatives market is becoming

---

[12] The quote has been taken from David Harvey's lecture „The Enigma of Capital", which he held at the London School of Economics on 26 April 2010. Available at: http://www2.lse.ac.uk/public Events/events/2010/20100426t1830vOT.aspx#generated-subheading2 (last accessed 29 June 2010).

increasingly more complex. This is expressed, for example, in the inclusion of carbon in so-called „commodity index funds" (Suppan 2009).

The necessity of a European „subprime carbon" (FoE 2009) cannot necessarily be derived from these elements, but they do show us that the new market-based instrument „has both consolidated and blurred the boundaries of EU climate governance, creating multiple opportunities for industry to influence the design and operation of the EU ETS in ways that may ultimately undermine its governability" (Bailey/Maresh 2009: 458).

3.3 Nature as Robinsonade

The sellability of a right to damage nature contains yet another contradiction between the natural side and the exchange value of the new commodity, which is articulated most clearly in the linguistic-symbolic production of climate change. Sellability, i.e. transferability aims at the physical isolation and separability of a part, of a portion of nature from the context of a comprehensive ecosystem. Nature – that is the third element of valorisation – is individualised. It becomes the protagonist of its own Robinsonade. Entities are constructed which do not exist in that form in the interdependent external nature of man. Fishing certificates, for example, assess the value of a fish independently of the water in which it swims. The logic of monetarisation expresses an atomistic point of view. The economic value of entire ecosystems results from the sum of its artificially separated individual parts, each of which would have no chance of survival on its own. „This way of pursuing monetary valuation tends to break down when we view the environment as being construed organically, ecosystemically, dialectically (…) rather than as a Cartesian machine with replaceable parts. Indeed, pursuit of monetary valuation commits us to a thoroughly Cartesian-Newtonian-Lockeian and in some respects ‚anti-ecological' ontology of how the natural world is constructed." (Harvey 1996: 153)

This anti-ecological ontology leads to society's search for a new hegemonial pattern of the appropriation of nature being characterised by a tendency to negate the immance of society and nature. The symbolic-linguistic production of climate change is reduced to the isolated consideration of greenhouse gas emissions and emission reduction corridors (the EU uses the catchy formula: 20% fewer emissions by 2020), so that societal relationships with nature are reflected back to the actors as a natural characteristic of the commodity carbon. Technological paradigms and social relationships disappear from focus – CO_2 becomes a fetish (Lohmann 2010). The understanding of a complex problem relating to society's interchange of materials with nature can in this way be rationally portioned. In the halls of the „epistemic communities" it is transformed into a point on the agenda which can be comprehended and dealt with managerially by the „Berufsmensch" (cf. Swyngedouw 2009: 379ff.; Crouch 2008). The fetishised forms of market, efficiency and technology become the problem solvers. The addressee is an abstract humanity which sees itself faced as a collective with one and the same threat. With the effect that issues such as the major role of fossil energy sources for economic growth, the historical responsibility of accumulated greenhouse gas emissions, the question of justice arising from the different effects of climate change etc. are excluded from the linguistic symbolisation. The social forms which have been brought forth by climate change can be modified in this way so that the societal relationships of power which are associated with them remain untouched. Swyngedouw's (2009: 378) normative exhortation, „to proceed to direct political and social discourse over how the socio-ecological coordinates of our everyday life, the production of new socio-natural configurations and the organisation of social metabolism (which is usually known as capitalism) in which we live should be reorganised", cannot be honoured on the basis of the „strategic selectivity" characteristic of the managerial discourse. The managerial discourse, rather, is like the attempt to produce the intractable first nature linguistically in such a way that it can be reconciled with the renewed second nature. It remains questionable here, to what extent the symbolically-linguistically initiated re-accentuation of the material production of nature will have to be transformed sooner or later into a ‚third or fourth reflection' of the forms of the human appropriation of nature. After all, the „abstractly identifying reasoning" which is bound up with the exploitation and the „domination of nature" (Görg 2008: 478f.) tends in its fetishised form to deny the non-identity of nature.

3.4 Qualitative indistinguishability

This does not of course remain without consequence for the material practice of ‚Robinsoning'. After all, the individualisation of the commodity nature is made possible by the qualitative homogenisation of disparate phenomena. The particularity of a seed, of a genome, of a tree or a fish can only be „internalised" into economic activity if it is monetarily categorised as an atomised ‚service' and its biophysical uniqueness is ignored (Robertson 2004). Monetarisation requires – that is the fourth point – abstraction: abstraction in a twofold sense (Castree 2003: 281). On the one hand we have to functionally abstract from the fact that qualitatively different entities are reduced to a uniform measure. Only their similarity or comparability is classified. The relevance of this aspect is directly observable when we look at the different greenhouse gases, which for the purposes of trading all have to be broken down into „CO_2 equivalents" with commensurable „global warming potential" (GWP). Each greenhouse gas has a different effect on the climate with regard to intensity and duration (Lohmann 2006: 100; McKenzie 2008). In contradiction of the facts, for the purposes of trade it must be assumed that there is a clear monetary, i.e. quantitative relationship between the greenhouse gases – an unequivocal relationship which science cannot confirm. Science, politically represented by the Intergovernmental Panel on Climate Change (IPCC), operates on the basis of probabilities. Once these are known they characterise the activity of the market subjects by giving it structure. Thus, the IPCC agreed at the turn of the year 1995/96 that the greenhouse gas HFC-23 (emissions of HFC-23 occur in the production of the cooling agent HCFC-22) had a GWP of 11,700. Translated into the language of the market this means that the avoidance of one tonne of HFC-23 in China generates the additional right to 11,700 tonnes of CO_2 emissions in the EU.[13] Since HFC-23 can be burned with relatively little technical and monetary cost, this practice was made good use of. HFC-23 projects gave rise to 67% of all the *Certified Emission Reductions* (CERs) within the framework of the CDM in 2005 and 34% in 2006 (Capoor/Ambrosi 2007: 27).

The abstraction imperative of valorisation goes further, however. In order to turn carbon into a commodity the way in which the emissions of greenhouse gases are reduced must be abstracted from (Robertson 2004), since from the point of view of the exchange value it makes no difference whatsoever whether the certificated allowance has found its way onto the stock exchange via a slump in the economy, the „fuel-switch" from coal to gas or the switch to regenerative energies. In 2009, for example, the energy producer RWE reduced its emission profile by 17%, in the steel branch the figure was 33% and for the entire EU ETS 11% (Capoor/Ambrosi 2010: 21). In the EU ETS the plant owners are rewarded for these one-off effects resulting from business trends in that they can either redeem their accumulated superfluous certificates at a later point in time in the Kyoto Phase (2008-2012) or even carry them over into the next (2012-2020) trading period (*banking*). The qualitative indistinguishability of different technological paradigms and social relationships on the exchange value side therefore has the effect of preserving structure. Instead of substituting the fossilistic „created ecosystems", the emissions saved due to business trends today are burned in the next boom.

From the point of view of the material side this is precisely the decisive point: qualitatively it makes a significant difference whether greenhouse gas

---

[13] The EU ETS is linked to the *Clean Development Mechanism* (CDM) of the Kyoto Protocol by the „Linking Directive". The *Certified Emission Reductions* (CERs) which are created within the framework of the CDM are therefore convertible into the EUAs of the EU ETS.

emissions are avoided due to a variation in the fossilistic mode of development or because of a structural change towards a solar economy. Herein lie the limits to the strategy of ecological modernisation and the selective valorisation of external nature, as climate change is a „wicked problem" which „challenges established social values and institutional frameworks, defies analysis, and has no obvious solution" (Jordan et al. 2010: 4).

The solution is also so difficult because form and material, exchange value and use value of societal development cannot be separated from one another. Climate change is embedded in the *articulated* whole of the capitalist mode of production. It is therefore of central importance to first make a definition of the problem dimension which the co-production and co-evolution of the internal and external nature of man may be capable of determining. This could help us to avoid supporting a „reflexively broken strategy of the domination of nature" (Görg 2003: 130), as is expressed in the objectivisation of nature through the fetishised forms of utilisation.

4   Own times in economics, politics and nature

To summarise: the strategy of the valorisation of nature describes a qualitatively new and ambivalent process. Qualitatively new is not so much that nature is turned into a commodity. Use values provided by nature (human labour power, mineral raw materials, fossil energy sources) have always been valorised in the history of capitalism in order to consume them productively and to realise through them a surplus value on the markets. What is new is that the under-produced elements of nature (to use O'Connor's handy formula) on which the capitalist growth machine feeds are included in the microeconomic calculation of the market actors. With this „real subsumption of nature under capital" (Smith 2006) nature in all of its aspects becomes a factor of production, is given a price and must be rationally consumed. This process is, however, ambivalent in that with the granting of (temporary) property rights *first* a social relationship of power comes into existence or is perpetuated; the monetary form impressed onto nature is *secondly* necessarily arbitrary and crisis-prone; the commodified individualisation of nature *thirdly* contains an „anti-ecological ontology"; and with the abstraction inherent in the exchange value *fourthly* qualitatively important differences are levelled out. Structural contradictions of the monetarisation of nature can therefore be named. However, this does not adequately explain exactly how these contradictions function. Whether they possibly make the socio-ecological degradation even worse or whether they in fact provide stimuli – within the capital form – for sustainable relations with the external nature of man, cannot be adequately answered at the level of a structure-theoretical analysis. In order to do so it would require on the one hand an extension of the study of the acting subjects and their concrete practices in the appropriation of nature. On the other hand the appropriation of nature must be embedded in the structural forms of societal development.

Using the example of the EU ETS it has been shown that different capital factions struggle over the concrete shape of the market design in alliance with state and civil society forces[14]. This unfolds itself as a hybrid regulatory framework. This hotly contested process of market creation is over-determined by the „capital absorption problem" in a global debt economy. The global debt economy is primarily aimed at translating differences in the socially necessary labour time in the here-and-now into extra profits. With the existence of carbon as a commodity which can be traded on the financial markets an instrument is favoured which is subject to the structural constraint of an economic regime in which the present time has become „omnipresent" (Altvater 1997: 442). The following problem arises from this: how can the present and short-term calculation of the market subjects be successfully regulated so that it can be assembled to a long-term strategy of the reduction of accumulated atmospheric greenhouse gases? How can the *desynchronised* time-structure (cf. Rosa 2005: 96) of politics, economics and nature be coordinated in their temporal heterogeneity? This necessarily lands us in political logic, i.e. the strategic selectivity of the state. We obviously find ourselves in a dilemma here: if the state is to serve capital accumulation it must be able to adapt the temporal horizon of the capital form. Economic competitiveness and innovations accompany the domination of the faster. „Chronopolitics" (Virilio 1980 quoted in ibid.: 36) becomes a keyword of the accelerated modernity. This undermines the „time sovereignty" and the legitimatory basis of the state, however (Jessop 2002: 109). Political processes of discussion, negotiation and implementation are time-consuming and are subject to their own acceleration-resistant time. The dilemma can thus be formulated more precisely as follows: the state is the decisive political instance which could, in a second reflection process, use its influence to *resynchronise* the relationship of spiral-shaped economic and irreversible ecological time. In its strategic selectivity, however, it tends to be committed to growth and thus works towards a depoliticising (but possibly also towards a de-legitimising) of the omnipresent present time.

For the future examination of carbon markets it therefore seems imperative to place a stronger accent on the temporal structures of socio-ecological development in addition to the crossing of different spatial scales, as these create a

---

[14] To put it simply, we can differentiate between a ‚fossilistic bloc' of Fordist industries and national governments and an ‚ecological modernisation bloc', which unites the financial sector, the EU Commission and the NGOs of the „Climate Action Network Europe" (CNE).

sensibility for the precarious dimension of the strategy of „ecological modernisation": the capitalist growth machine should be decarbonised with the means of
western rationality and the acceleration cycle continue to turn untouched by this.

References

Altvater, E. (1992): Die Zukunft des Marktes. Ein Essay über die Regulation von Geld
    und Natur nach dem Scheitern des „real existierenden" Sozialismus. Münster: Ver-
    lag Westfälisches Dampfboot.
Altvater, E. (1997): Entbettung. In: Historisch Kritisches Wörterbuch des Marxismus.
    Vol. 3. Hamburg: Argument Verlag: pp. 438-444.
Altvater, E. (2006): Das Ende des Kapitalismus wie wir ihn kennen. Eine radikale
    Kapitalismuskritik. Münster: Verlag Westfälisches Dampfboot.
Altvater, E. (2006): The social and natural environment of fossil capitalism. In: Panitch,
    L.; Leys, C. (eds.): Coming to Terms with Nature. Socialist Register 2007. Mon-
    mouth: Monthly Review Press: pp. 37-60.
Bailey, I.; Maresh, S. (2009): Scales and Networks of Neoliberal Climate Governance: the
    Regulatory and Territorial Logics of European Union Emissions Trading. In: Trans-
    action of the Institute of British Geographers, Volume 34, Number 4, October 2009:
    pp. 445-461.
Bakker, K. (2005): Neoliberalizing Nature? Market Environmentalism in Water Supply in
    England and Wales. In: Annals of Association of American Geographers 95: pp.
    542-565.
Brand, U.; Görg, C. (2003): Postfordistische Naturverhältnisse. Konflikte um genetische
    Ressourcen und die Internationalisierung des Staates. Münster: Verlag Westfälisches
    Dampfboot.
Bumpus, A.; Liverman, D. (2008): Accumulation by Decarbonization and the Governance
    of Carbon Offsets. In: Economic Geography, Vol. 84, No. 2: pp. 127-155.
Capoor, K.; Ambrosi, P. (2007): State and Trends of the Carbon Market 2007, Washing-
    ton DC: World Bank.
Capoor, K.; Ambrosi, P. (2010): State and Trends of the Carbon Market 2010, Washing-
    ton DC: World Bank.
Castree, N. (2002): False Antithesis? Marxism, Nature and Actor-Networks. In: Antipode,
    Vol. 34, No. 1: pp. 111-146.
Castree, N. (2003): Commodifying what nature? In: Progress in Human Geography 27, 2:
    pp. 273-292.
Castree, Noel (2008): Neo-liberalising nature: the logics of de- and re-regulation. In:
    Environment and Planning A, Vol. 40: pp. 131-152.
Convery, F.; Redmond, L. (2007): Market and Price Developments in the European Union
    Emissions Trading Scheme. In: Review of Environmental Economic and Policy 1:
    pp. 88-111.
Crouch, C. (2008): Postdemokratie. Frankfurt am Main: Suhrkamp Verlag.

Ellerman, D.; Convery, F.J.; de Perthuis, C. (eds.) (2010): Pricing Carbon. The European Union Emissions Trading Scheme. Cambridge: Cambridge University Press.

Europäische Kommisssion (2008): Questions and Answers on the revised EU ETS. Brussels.

Friends of the Earth (2009): Sub-Prime Carbon: Re-thinking the world's largest new derivatives market. Washington.

Friends of the Earth (2010): Ten Ways to Game the Carbon Market. Washington.

Gilbertson, T.; Reyes, O. (2009): Carbon Trading: how it works and why it fails. In: Critical currents No. 7, November, Uppsala: Dag Hammarskjöld Foundation.

Görg, C. (2003): Nichtidentität und Kritik. Zum Problem der Gestaltung der Naturverhältnisse. In: Böhme, G.; Manzei, A. (eds.): Kritische Theorie der Technik und der Natur. Munich: Wilhelm Fink Verlag: pp. 113-133.

Görg, C. (2008): Gesellschaftliche Naturverhältnisse. In: Peripherie No. 112, Vol. 28: pp. 477-479.

Gramsci, A. (1991ff.): Gefängnishefte. Kritische Gesamtausgabe. ed. by Bochmann, K.; Haug, W.F. Hamburg: Argument Verlag.

Hardin, G. (1968): The Tragedy of the Commons. In: Science, Vol. 162, No. 3859: pp. 1243-1248.

Harvey, D. (1996): Justice, Nature and the Geography of Difference. Malden, MA: Blackwell Publishers.

Harvey, D. (2010): The Enigma of Capital and the Crisis of Capitalism. London: Profile Books.

Harvey, D. (2010): The Enigma of Capital. Lecture at the London School of Economics, 26 April 2010. Available at http://www2.lse.ac.uk/publicEvents/events/2010/20100426t1830vOT.aspx#generated-subheading2, last accessed June, 29th 2010.

Hauff, V. (1987): Unsere gemeinsame Zukunft. Der Brundtland-Bericht der Weltkommission für Umwelt und Entwicklung. Greven: Eggenkamp Verlag.

Jessop, B. (1990): State Theory: Putting the Capitalist State in Its Place. University Park, Pennsylvania: The Pennsylvania State University Press.

Jessop, B. (2002): Time and Space in the Globalization of Capital and Their Implications for State Power. In: Rethinking Marxism, Vol. 14, No. 1: pp. 97-117.

Jessop, B. (2008): State Power. Cambridge: Polity Press.

Jordan, A.; Huitema, D.; van Asselt, H.; Rayner, T.; Berkhout, F. (eds.) (2010): Climate Change Policy in the European Union. Confronting the Dilemmas of Mitigation and Adaptation? Cambridge: Cambridge University Press.

Kanter, J. (2007): In London's Financial World, Carbon Trading is the New Big Thing. New York Times, July 6th.

Latour, B. (1993): We Have Never Been Modern. Cambridge Mass.: Harvard University Press.

Lohmann, L. (2006): Carbon Trading: A Critical Conversation on Climate Change, Privatisation and Power. In: Development Dialogue No. 48, September. Uddevalla: Dag Hammarskjöld Foundation.

Lohmann, L. (2009): Climate as Investment. In: Development and Change. Vol. 40, No. 6: pp. 1063-1083.

Lohmann, L. (2010): Commodity fetishism in climate science and policy. Presentation to Imperial College London, February.

MacKenzie, D. (2008): Making Things the Same: Gases, Emission Rights and the Politics of the Carbon Markets. Edinburgh: University of Edinburgh.

Maddison, A. (2001): The World Economy. A Millenial Perspective. Paris: OECD.

Mansfield, B. (2004): Neoliberalism and the Ocean: „Rationalization", Property Rights, and the Common Question. In: Geoforum 35: pp. 313-326.

Marx, K. (1961): Capital, Vol. I, Moscow (translated from the third German edition by Samuel Moore and Edward Aveling and edited by Frederick Engels).

McAfee, K. (1999): Selling Nature to Save it? Biodiversity and Green Developmentism. In: Environment and Planning D: Society and Space 17: pp. 133-154.

McKinsey & Company (2009): Pathways for a low-carbon economy.

Meadows, D. et al. (1972): The Limits to Growth. New York: Universe Books.

Mol, A. (2003): Globalization and Environmental Reform: The Ecological Modernization of Economy. Cambridge Mass.: MIT Press.

Morton, T. (2007): Ecology Without Nature. Cambridge Mass.: Harvard University Press.

Newell, P.; Paterson, M. (1998): A climate for business: Global warming, the state and capital. In: Review of International Political Economy Vol.5, No.4: pp. 679-704.

Newell, P.; Paterson, M. (2009): The politics of the carbon economy. In: Boykoff, M. (ed.): The Politics of Climate Change: A Survey. London: Routledge: pp. 80-99.

Newell, P.; Paterson, M. (2010): Climate Capitalism: Global Warming and the Transformation of the Global Economy. Cambridge: Cambridge University Press.

Oberthür, S.; Tänzler, D. (2007): Climate Policies in the European Union: The Influence of International Regimes in Policy Diffusion. In: Harris, P. (ed.): Europe and Global Climate Change: Politics, Foreign Policy, and Regional Cooperation Cheltenham: Edward Elgar: pp. 255-278.

O'Connor, J. (1988): Capitalism, Nature, Socialism. A Theoretical Introduction. In: Capitalism Nature Socialism, Vol. I, No. 1: pp. 11-38.

Panayotakis, C. (2006): Working More, Selling More, Consuming More: Capitalism's ,Third Contradiction'. In: Panitch, L.; Leys, C. (eds.): Coming to Terms with Nature. Socialist Register 2007. Monmouth: Monthly Review Press: pp. 254-273.

Pearce, D.; Markandya, A.; Barbier, E. (1989): Blueprint for a Green Economy. London: Earthscan Publications.

Peck, J.; Tickel, A. (2002): Neoliberalizing Space. In: Antipode 34: pp. 380-404.

Point Carbon (2004): What Determines the Price of Carbon? Carbon Market Analyst, Point Carbon, 14 October 2004.

Robertson, M. (2004): The Neoliberalization of Ecosystem Services: Wetland Mitigation Banking and Problems in Environmental Governance. In: Geoforum 35: pp. 361-373.

Rosa, H. (2005): Beschleunigung. Die Veränderung der Zeitstruktur in der Moderne. Frankfurt am Main: Suhrkamp Verlag.

Skjaerseth, J.B.; Wettestad, J. (2008): EU Emissions Trading. Initiation, Decision-making and Implementation. Farnham/Burlington: Ashgate.

Smith, N. (1996): The Production of Nature. In: Bird, J.; Putman, T.; Robertson, G.; Tickner, L. (eds.): Futurenatural. London: Routledge: pp. 35-54.

Smith, N. (2006): Nature as accumulation strategy. In: Panitch, L.; Leys, C. (eds.): Coming to Terms with Nature. Socialist Register 2007. Monmouth: Monthly Review Press: pp. 16-37.

Smith, N. (2008): Zur kapitalistischen Produktion von Natur. In: Das Argument 279: pp. 873-878.

Stern, N. (2007) The Stern Review: The Economics of Climate Change. Cambridge Mass.: Perseus Books.

Strange, S. (1994): States and Markets. London/New York: Pinter.

Suppan, S. (2009): Speculating on Carbon: The Next Toxic Asset. Institute for Agriculture and Trade Policy. Minneapolis.

Swyngedouw, E. (2009): Immer Ärger mit der Natur. „Ökologie als neues Opium für's Volk". In: Prokla, issue 156, Vol. 39, No.3: pp. 371-389.

van Asselt, H. (2010): Emissions trading: the enthusiastic adoption of an ‚alien' instrument. In: Jordan, A.; Huitema, D.; van Asselt, H.; Rayner, T.; Berkhout, F. (eds.): Climate Change Policy in the European Union. Confronting the Dilemmas of Mitigation and Adaptation? Cambridge: Cambridge University Press: pp. 125-144.

Weale, A. (1992): The New Politics of Pollution. Manchester: Manchester University Press.

Weber, M. (1973): Vom inneren Beruf zur Wissenschaft. In: Weber, M.: Soziologie – Universalgeschichtliche Analysen – Politik. Stuttgart: Kröner.

Weber, M. (1980): Wirtschaft und Gesellschaft. Tübingen: Mohr.

Weber, M. (2006): Die protestantische Ethik und der Geist des Kapitalismus. Munich: Verlag C.H. Beck.

Economic Growth and Climate Change: Cap-And-Trade or Emission Tax?

Edward Nell / Willi Semmler / Armon Rezai

1 Introduction

Economic growth and the globalization of goods and capital flows led to an unsustainable level in the consumption of natural resources. This entailed a steady increase of pollution and climate change. The work of the Intergovernmental Panel of Climate Change (IPCC) was crucial in the dissemination of these findings and in the discussion leading to the Kyoto Protocol (IPCC 2006, 2007). One of the protocol's controversial aspects was the choice of policy instruments in the curbing of carbon emissions. Generally speaking, two approaches seem to be favored: „cap-and-trade" implying a trading of emission rights and „carbon tax" implying a taxation of emissions.

Many of the debate's protagonists protagonists are evoking economic theory to offer guidance and support for their arguments. Our chapter tries to shed some light on the various theories, which attempt to explain the relationship between economic growth and climate change (especially Stern 2007, Nordhaus 2008, and Uzawa 2003), and to evaluate them critically along the ideas of intergenerational equity and economic externalities.

Given these theories, we want to explore what they have to say about the policy instruments of cap-and-trade schemes and a carbon tax. In line with recent IPCC reports, we argue that taxes are to be preferred over market oriented schemes as these are subject to many drawbacks such as their complicated institutional structure and implementation and their price volatility. In particular, given the recent stock market meltdown and volatility, the cap-and trade system presumably will not be very effective in providing sufficient incentives for reducing CO_2 emissions.

The chapter has the following structure: in section 2 we discuss some empirical facts regarding the relationship, causes, and consequences of global warming. In section 3 we discuss economic theories of intergenerational equity and climate change and propose a multiple equilibria model of the interaction of growth and climate change. In section 4 this is followed by a historical investigation into the theories of economic externalities of Pigou and Coase. In section 5 we analyze which of the policy instruments mentioned above are preferable in the light of these theories. Section 6 presents the European experience with cap-and-trade schemes.

2   Growth and Climate Change: Empirical Facts

The globalization of economic activities since the 1980s and 1990s, accelerated through free trade agreements, liberalized capital markets and labor mobility, has, since the 1970s already, brought into focus the issues related to global growth, resources and environment. The industrialization in many countries during the last 100 years or more and the resource based industrial activities have exhausted resources, mostly produced by poor and less developed countries. The tremendous industrial growth in the world economy, in particular since World War II and strong economic growth in the last two decades in some regions of the world, for example in the US, Asia and some Latin American countries, have generated a high demand for specific inputs. Renewable as well as non-renewable resources had been in excess demand and they are threatened to be depleted. In particular the growing international demand for metals and energy derived from fossil fuels, as well as other natural resources, which are often extracted from developing countries, has significantly reduced the years until exhaustion of those resources. The International Energy Agency estimates a further depletion of resources and price increases in the long run.

On the other hand the industrialization and resource based activities has had strong external effects by polluting and degrading the environment. Not only does the environmental pollution strongly affect current generation, but the environmental degradation affects also future generations. Since the conference and protocol of Kyoto in 1997 the global change of the climate has become an important issue for academics as well as for politicians. Both the overuse of resources as well as the environmental pollution and climate change has brought into focus the issue of inequity.

Since the creation of the Kyoto protocol, which was ratified by more than 170 countries in the meantime, the topic of global climate change has gained growing attention from academics and politicians. Especially within the academic realm of economics, the aspect of intergenerational inequity was brought into focus. Basically global warming represents an unequal treatment of individuals across generations. Current generations extensively reap the gains of energy-intensive production which entails permanent damage to the environment and climate change while future generations have to bear the costs of these negative externalities in form of lower quality of live and bio-diversity.

The role of such imperfect linkages between generations in economic growth and Greenhouse Gas (GHG) emissions has become a major topic in economic research, as for example in the works of Uzawa (2003), Nordhaus (2008), Stern (2007), Heal (1998), Greiner and Semmler (2008) and the various reports of the IPCC (2006/7). The greatest concern with this respect in their work is CO_2 emissions. Since the beginning of the 19th century yearly emissions have quadrupled with a 70% increase occurring within the past 30 years. The concentration of CO_2 in the atmosphere rose from 280 ppm in the year 1750 to 379 in the year 2005. At the same time, a warming of the global climate is unequivocal and can be observed in the increases in global average air and ocean temperatures. Since 1900 global temperature has increased by roughly 0.7 degree Celsius. Figure 1 shows the global development of atmospheric CO_2 since 1750. The effects of such changes are drastic: Polar ice caps have been melting causing upswell of the global mean sea level while the likelihood and severance of extreme weather events increased.

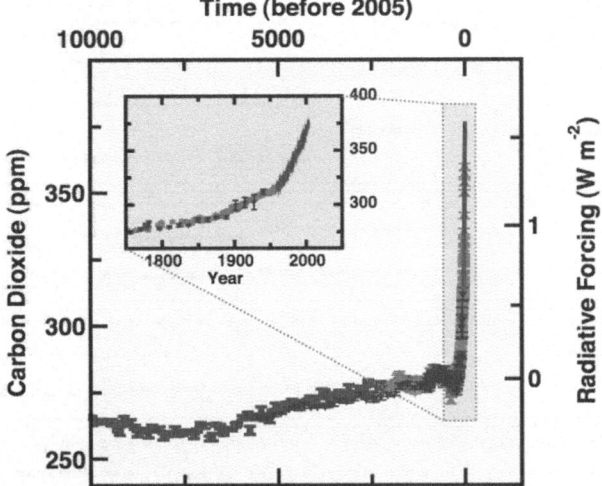

Figure 1:   Atmospheric CO_2 over the last 10,000 years and since 1750,
             Source: IPCC (2007, p. 38)

Given current developments, scientific forecasts predict a likely temperature rise at the end of the century by 2-4 degree Celsius, sea levels are likely to rise by 28 – 43 cm, the Arctic sea ice is likely to disappear (in the second half of the century), and the probability of large parts of the world experiencing an increase in the number of heat waves, droughts, intensified tropical storms, and floods is high.

In fact, there are two types of inequities. First, the overwhelming fraction of exhaustible resources, located in the South, are exhausted by the industrialized countries of the North, which have become the main polluters of global environment by causing global warming. The second type of inequity results from the unequal treatment across generations. Current generations extensively use up resources, pollute the environment and cause global climate change. This produces negative externalities for future generations and future generations are treated unequally.

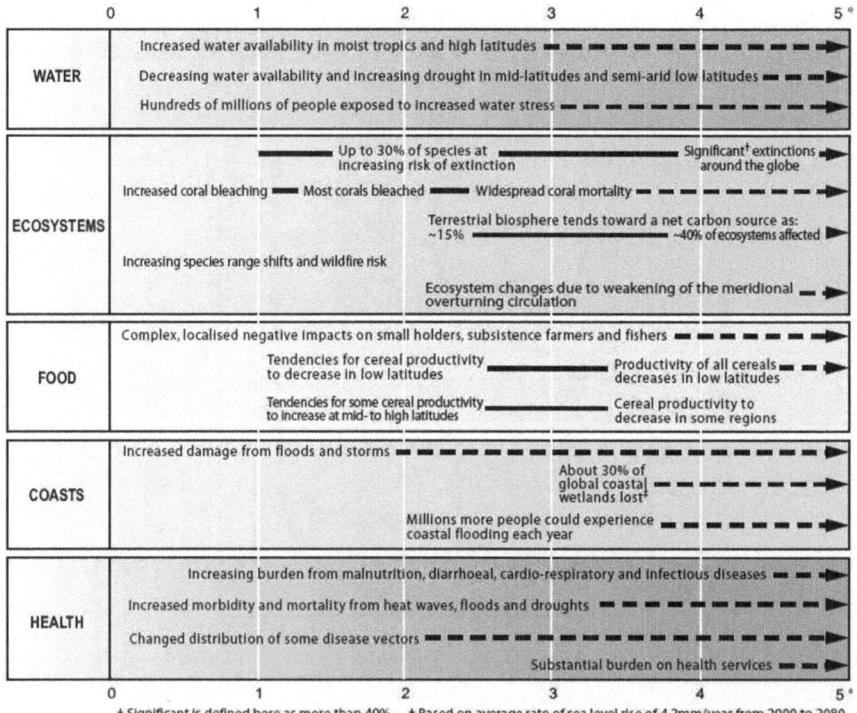

Table 1: Examples of impacts associated with global average temperature change, Source: IPCC (2007, p.51)

3   Growth and Climate Change: Theory and Modeling

Impacts of climate change will vary regionally but, aggregated and *discounted* to the present, they are likely to impose considerable future costs which will increase over time as global temperature increases. An extensive academic literature explores the issue of how to evaluate future damages as compared to current cost to avoid them. As pointed out above, due to the intergenerational aspect of the problem, preventing damages in the future requires efforts today. One direction to address this issue in economics is to employ the instruments of welfare analysis which weight future damages against their current costs. The crucial issue here is the discounting of the future: How much effort (current cost of change) should be undertaken depends on how we evaluate (*discount*) future damages. The degree of discounting is clearly a key in such an undertaking and varies widely in different studies. While Stern (2007) uses a very low discount rate of 1.4%, Nordhaus (2008) proposes to discount future welfare and damages by a discount rate of 5%. Weitzman (2008) proposes a discount rate of about 4%. In academia there exists a long standing debate on how exactly the welfare of future generations should be taken into account. While alternative approaches have been developed for valuating the future, see e.g. Heal (1998), overall there seems to be consensus that there exists a (rising) cost of future damages due to climate change. The calculation of such costs is yet more complex because potential catastrophe losses, response lags, the treatment of risk, and non-economic impacts have to be included in the analysis.

Yet, overall there are important feedback effects: as mitigation efforts (depending on the discount rate) and growth affect climate change, so would also future growth depend on climate change.

A typical model where this interdependence of growth and climate change is treated in stylized way is the model by Greiner and Semmler (2008). Crucial for this approach is the albedo which is the fraction of incoming energy (energy from the sun) that is reflected back. On the other hand (1-albedo) is the fraction absorbed by the earth. As figure 2 shows the latter fraction rises with the temperature T (measured in Kelvin). This is, for example, due to the melting of arctic ice cap (for further positive feedback effects between current and future temperature, see Lovelock, 2006).

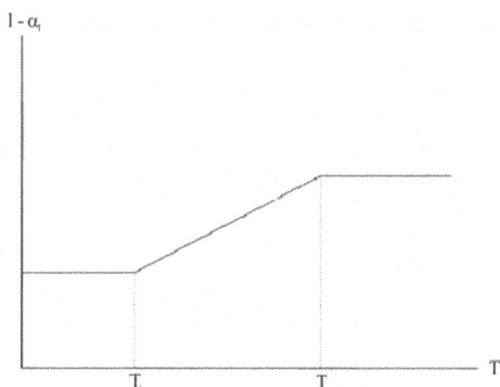

Figure 2:   Absorbed Solar Energy (1-α) as a Function of Global Temperature,
Source: Greiner and Semmler (2008)

The model by Greiner and Semmler (2008) shows that there are three possible long run growth rates associated with different temperature. The normal temperature measured in Kelvin, is T=288. This corresponds to 15 degree Celsius. The three possible growth rates are:

Scenario	1-α	Temp.	GDP Growth
I	0.2093	3.7°C ↑	2.6%
II		4.5°C ↑	2.3%
III	0.2171	6.2°C ↑	2.2%

Table 2:   Results for Different Albedo Assumptions,
Source: Greiner and Semmler (2008)

Table 2 illustrates the three different scenarios investigated in Semmler and Greiner. A stability analysis shows that only regime I and III are stable. Regime II is unstable and will not be realized in reality. The critical value of the middle case demonstrates the inherent effects of nonlinearity on the stability of possible growth paths. A small deviation around the lower equilibrium I causes no long run effects. If the critical value, which separates the two centers of attraction, is transgressed, however, there will be significant climate change with irreversible effects.

This means that if countries postpone the reduction of CO_2 emissions, temperature through the albedo feedback may rise so much that only the low growth – high temperature growth path can be run. This makes climate policy urgent. The implementation of carbon reducing efforts by the governments into such a model with multiple equilibria shows that these efforts not only lower equilibrium temperature levels, but also increase economic growth. Thus the economy will be put on a superior growth path. The reduction of carbon emission could come from a carbon tax. The logic is that a higher emission tax will induce greater private abatement efforts.

Tax Level	Temp.	GDP Growth
0.11%	3.4°C ↑	2.60%
0.40%	0°C	2.80%
0.55%	0.8°C ↓	2.77%

Table 3:   Effects of a carbon tax, Source: Greiner and Semmler (2008)

The considerations above have strong implications on the controversy about discount rates: Discount rates are not as important as the danger of doing too little too late. Before we discuss this matter of an emission tax further, we want to give an overview on recent discussions on mitigation methods which were initiated by ICPP (2007).

4   The History of Regulating Economic Externalities: Pigou versus Coase

Uzawa (2003), Nordhaus (2008), and Greiner and Semmler (2008) suggest emissions taxes to reduce emissions of greenhouse gases. This idea of regulating economic externalities originates in Pigou (1932). Other economists follow a market-oriented approach based on Coase (1960) who suggested creating and trading emission rights in order to internalize an economic externality. In this section we give a brief overview on the long standing controversy between those two approaches and evaluate their applicability to global warming.

Marshall is usually associated with the development of supply and demand and their clearing in markets. Soon after his theorizing, however, it was discovered that there exist cases where a substantial economic impact, positive or negative, was delivered through the market mechanism, but nevertheless lay in a sense outside of the process of supply and demand. That is, although economic agents were benefited or harmed in regard to their income or wealth, the actions that impacted them were not driven by the motive to do so, nor were the agents themselves obliged to pay for the benefits, or able to escape the harm by making payments.

Pigou, Marshall's successor in the Professorship at Cambridge, proposed in The Economics of Welfare to deal with this by a system of taxes and subsidies. Activities that led to negative externalities should be taxed; the proceeds could be devoted to compensating those injured, or rectifying the situation. But the tax itself would reduce the amount of the activity. Activities with positive externalities should be subsidized, thereby increasing the amount provided, so as to ensure that the full social benefits would be achieved – social marginal rewards should be equated to social marginal costs.[1] By doing so, an optimal position will be reached. The strength of this approach is that it reaches clear and definite conclusions about policy. Arguing from conventional assumptions it shows that government intervention can correct the problem of externalities.

However some important reservations should be noted. First, the approach is normally set out in the form of a static equilibrium model (as described in the footnote.) Modern economies are not static; both the economies and the problems are growing, and one major concern is that climate problems are growing faster than our economies. A static framework cannot deal with this. The fact that it is also an equilibrium framework just adds a further note of unreality. Actual growing economies are in flux; the composition of output and the distribution of income as well as technologies and organization of production are all changing. They are not in equilibrium, and the problems of emissions have to be dealt with in conditions that may include change and disorder.

It is not just that actual economies are dynamic, and seldom exhibit equilibrium; we know now that markets can be systematically unstable and volatile, often within definable limits, and they can be operating in ways that lead to dynamic patterns of change. Markets are not characterized by easily defined and unique stable equilibrium positions. They are more complex, and more fragile than their advocates have understood. They generate powerful forces for innovation, but for just that reason, their movements can be unsettling.

This leads to the second point: the Pigouvian framework is too simple in another way; the models can generally be solved pretty easily for unique solu-

---

[1] Economists following Pigou have set this out in simple neo-Classical models, in which consumer utility depends positively on output, negatively on pollution, output depends positively on inputs of labor and capital, but also on the production of waste (curbing waste will reduce output). (Output may also be reduced by pollution.) Pollution, in turn, depends positively on waste products. This gives rise to a simple maximizing problem with the usual first and second order conditions. The model can be extended to include defensive activities by households, trying at their own expense to defend themselves from pollution. Maximizing consumer utility will lead to the result that firms should produce to the point where the marginal product of their waste emissions just equals the marginal damages those waste emissions do to consumers. And *this* is the level of the Pigouvian tax or ‚effluent fee‘ that the government should impose directly on waste emissions. This tax will produce an optimal result. In this model there is no room for compenzation of victims.

tions. This is very unlikely in a dynamic framework, and it is probably a mistake to be looking for a unique policy that will provide the clearly defined ‚best' outcome. Instead we should expect finding a number of policies that will most likely improve the situation, bearing in mind that, usually, it will not be possible to declare one of them the ‚best'.

In contrast to the Pigouvian tradition Coase (1969) argued in „The Problem of Social Costs" that rather than intervening with taxes, governments working with the parties affected by externalities should find ways of changing liabilities and property rights, if possible in ways that will permit market incentives to drive the system to a superior position. Coase did not propose specific models, but his followers have developed them. ‚Cap-and-trade' is certainly in that tradition.

The idea is that by redefining property rights governments can make use of market forces to accomplish their goals, rather than establishing a bureaucracy to force businesses to follow costly rules. Governments must find ways to put businesses in charge for things which are currently considered outside their interests, such as air quality and climate change. Markets can be used to settle how to distribute the costs of reducing air pollution, or excess carbon emission, and the burdens of bearing it. The desired or acceptable level of pollution is decided by Government, and then ‚rights to pollute', adding up to that amount, are distributed among those who benefit from the activity causing the pollution. These rights can be bought and sold. A firm that needs to pollute more can buy the rights from firms that can easily reduce their pollution. If the neighbors of a polluter can't stand it, they can bid for that polluter's rights, thereby compensating the firm for the cost of reducing pollution. The idea is to keep government out of the details – bureaucrats are likely to tangle things in red tape. Instead of micromanaging, the government should create property and other rights that will give rise to tradable claims.

Coase starts from very realistic examples – a train running through farmland causing uncompensated damage, a noisy shop next to a doctor's office, cattle farmers and wheat farmers – in each of which the normal activity of one party damages the normal activity of the other. He carefully parses the ways in which rights might be assigned, determining who bears the burden of liability, comparing the possible economic results. It is clear that the set of possible outcomes is far larger than imagined by Pigou, but Coase also accepts a very simple and standard account of prices and markets – essentially Pigou's. They neither appreciate the importance of market dynamics, multiple equilibria and market instability, nor do they consider the problem of externalities in a growing economy.

Interestingly, Coase, unlike his pro-market followers, did not contend that a reassignment of rights was always the best way to go. He accepted that a system of taxes and subsidies might as well do the job (Coase, 1969). The more or less standard view of Coase's argument – surely at least partly justified – is that he proposed that we should always consider the possible ways to re-assign property rights in cases of externalities. If such reassignment could effectively internalize those externalities, the market could simply go ahead (in the US such a system reduced acid rain and basically brought about the removal of lead from gasoline, ‚Cap-and-Trade‘ proposes to do this for carbon). But there is more to Coase than that. First, he redefines ‚factors of production‘ as rights, rights to carry out certain kinds of activities. These rights can be redefined, and the question is, which bundle of rights, assigned which way, will be the best. He sees a close connection between technology and the legal system, a connection that can affect the direction of innovation. But he does not assume that the present arrangement is best, nor does he necessarily argue against government intervention or taxation. Re-assigning rights, redesigning the bundles, is most certainly intervention. His chief point is that there are many possibilities. He also recognizes in his examples, though he does not stress it in theoretical discussions, that the burden of liability – and of taxation – can also provide a stimulus to innovation. This point is neglected by many of his followers.

Regulation, and government intervention generally, is indeed very powerful and the impacts will set off responses, that will develop over time through market processes. In potentially unstable markets, and in dynamic conditions, the results may be very far from those desired.[2] Indeed in the case of ‚Cap-and-Trade‘ the chief effect of market instability and price volatility may be to create problems rather than to provide incentives to desirable behavior. It might be best to design a simpler system that would lead to market responses which would reliably bring about responses in the right directions; properly designed taxation can do this. Any taxes, even very light ones, will be likely to trigger responses. After all they affect the earning and spending of money. But if we think through the program carefully, we can develop a system of taxes that should generate desirable responses, moving the market in the directions we want, and provide incentives for innovating in the most desirable ways. As we shall see, the better and more reliable course is to depend on the strong and direct incentives created by taxes.

---

[2] We see this in the case of illegal drugs, for example. Illegalizing drugs has a very impact but not the desired one of eliminating the market. Instead it drives up the price and makes drugs more profitable for the successful cartels. Police work also tends to eliminate competition, and block entry, making life easier for the established ‚firms‘.

5   Cap-and-Trade or Carbon Taxes?

The most debated and important of these policy measures are emission trading and tax policies, best known as „cap-and-trade" and „carbon tax". For many economists (Uzawa, 2003, Nordhaus, 2008, Mankiw, 2007, IPCC, 2007) an emission – or carbon – tax is preferable to a cap-and-trade system. This is due to advantages of the former and shortcomings of the latter:

Tradable permits (Cap-and-trade system) require that the actual polluter can be identified, for example firms. Enforcement of the cap is difficult and trading of emission certificates is exposed to speculative investments, generating a high volatility of the carbon price as the European example shows, see section 7.

Carbon Tax, on the other hand, allows for a broader application, including energy supply, major polluting industries, the service sector, transport system and households. Furthermore, the generated tax revenue can be used to reduce other taxes and tax funds, to compensate developing economies, or to induce climate-friendly investment behavior (see Uzawa, 2003).

Although more and more economists seem to lean toward a carbon tax, policy makers appear to tend toward the market-based cap-and trade system. Overall, we obtain in our model similar results as in the growth literature regarding government actions. A zero emission tax is not necessarily welfare improving (Greiner and Semmler, 2008). As above mentioned permits are open to speculative investments and exhibit a high volatility of carbon price. On the other hand, a carbon tax allows for a broader application: energy supply, industry, service sector, transport system and households. The tax revenue can be used to reduce other taxes and tax funds to be used to compensate developing economies.

6   European Efforts and Experiences: The high Volatility of the Carbon Price

Europe was prior in the implementation of the Kyoto protocol. As an instrument for reducing GHG the European Union has implemented the cap-and-trade system since 2005. The experience of the cap-and-trade system in the EU showed that, in practice, the cap-and trade system does not work efficiently. In addition to the disadvantages of tradable permits mentioned above, poor implementation practices worsened the system's performance: permits were granted rather than auctioned off which entailed windfall profits for the holders, costs have been passed on, too many industries and activities have been left out, and sometimes the carbon price crashed, due to ill-timed permit issuance. The crucial appoint is the volatility of the carbon price.

Figure 3:   Eurostoxx (10/03/05 – 20/03/08, week days only),
               Source: Bloomberg

Figure 4:   Carbon Price €/t (10/03/05-20/03/08, week days only),
               Source: Bloomberg

Already before the introduction of emission trading several critics argued that the price per tonne CO_2 would be too volatile in order to send correct signals to market participants regarding the true cots of emissions and the appropriate level of abatement. Figure 3 and 4 show the development of the European stock market (represented by the Eurostoxx Index) since September 04 and the price of

CO_2 since September 2005 (the starting point of emission trading in Europe). Both time series end on March 20th, 2008.

Mere inspection of both graphs shows that the carbon price is much more volatile and unstable than the stock market index Eurostoxx. To get a more precise picture of the volatility, table 4 summarizes the standard deviations for various real and financial variables in Euroland. The standard deviation of employment is the lowest for employment with 0.32% followed by consumption and GDP with volatility in the 0.60%s. Return on short term government debt has a standard deviation of 0.89% and investment fluctuates with 1.4% of its mean. Return on equity (in our case Eurostoxx) has a volatility which is more than 10 times GDP's with 3%. The standard deviation of the return on the emission price is much higher with 11% for the forward and more than 10 times the one of equity with 37% for the spot price.

The high volatility on the return, due to the tremendous fluctuations of the emission price, poses considerable uncertainty for firms in their investment decisions. Moreover, when the carbon price becomes too low, efforts to reduce CO_2 will rapidly fall.

Variable	Standard Deviation (in %)
Employment	0.32%
Consumption	0.61%
GDP	0.65%
Return on Government Debt (Short term)	0.89%
Investment	1.4%
Return on Equity	3%
Return on Emission Price (Forward)	11%
Return on Emission Price (Spot)	37%

Table 4:   Volatility of selected variables (quarterly data for EU countries from 1970-1997), Source: Data on Emission Price Bloomberg, all other Semmler (2006)

Recently new proposals to further reductions in GHG have been passed (EU decisions, January and December, 2008). The above-mentioned further reduction aim in the new EU guidelines strive for a reduction of CO_2 emission by 20% from their 1990 level by 2020 and an increase in alternative energy usage to 20% of total energy supply. While these aims are ambitious, they are still mostly pursued through the permits system (though the introduction of an auction system is now planned). On the other hand, the EU is also aiming to reduce CO_2 through a carbon tax, in particular on imported goods. Overall, given the European experience, the carbon tax appears to be a better solution as the above volatility study has shown.

7   Conclusions

Given the recently published scientific evidence on global warming and its damages, such as the numerous reports by the IPCC, the importance of climate mitigation has been sufficiently demonstrated. The urgency on climate actions becomes particularly true if the positive feedback effect of „temperature on temperature" through the endogenous albedo effect holds true. This produces the danger of doing „too little too late." Our multiple equilibria model motivates this urgency. As shown, our considerations have strong implications on the controversy about discount rates: Discount rates are not as important as the danger of delaying actions. The last report by the IPCC (4th Assessment Report) has urgently suggested a broad range of mitigation policy measures, such as integrating climate policies, broader development policies, regulations and standards, financial incentives, voluntary agreements, and information instruments to control and reduce Greenhouse Gas (GHG) emissions.

It also emphasizes the role of technology policies to achieve lower CO_2 stabilization levels, a greater need for more efficient R&D efforts, and higher investment in new technologies during the next few decades (for achieving stabilization and reducing costs). Further recommendations include government initiatives for funding or subsidizing alternative energy sources (solar energy, ocean power, windmills, biomass, and nuclear fusion).

Moreover, it is necessary, as the IPCC requests, to integrate climate policies in broader development policies, regulations and standards, for example taxes and charges, financial incentives, voluntary agreements and information instruments. Overall the IPCC stresses the fact that the effectiveness of such policies ultimately depends on national circumstances, their design, interaction, stringency and implementation. Yet, the major instruments that the IPCC and numerous well-respected economists propose are two specific tools to reduce GHG in order to fulfill the agreements of the Kyoto protocol. These two tools – decentralized market trading of emission right and carbon taxation – were the subject of our article. Both measures have a long standing history in economic theory originating in the works of Pigou and Coase.

The experience of the European Union with emission trading and recent economic models demonstrate that there is a clear advantage in choosing carbon taxes over carbon trading schemes. The advantages are the universal applicability, better efficacy, and lower set-up costs due to existing administrative institutions. The disadvantages of carbon trading are the arbitrary distribution of rights to special interest groups (rather than auctioning them off) and the disproportionate volatility of the emission price due to uncertainty in the overall quota and financial speculation. Hence, revising the existing, ill-conceived trading scheme in Europe and the introduction of carbon taxes in the developed countries are necessary conditions for meeting the commitments of the Kyoto protocol and the ambitious goals of the European Union and for ultimately obviating further increases in CO_2 emissions.

For the US it seems to hold that the effort to be undertaken to reduce CO_2 emissions should be even higher than for Europe. The US had much higher growth rates than Europe since 1990 and lower energy efficiency improvements, and thus higher CO_2 emissions than Europe. The reduction by 20% of CO_2 emissions as compared to the 1990 level (as required by the Kyoto protocol) now means effectively a reduction of the CO_2 by more than 30% of the 1990 level, according to some experts. In which steps this will be achieved by the Obama administration and the new congress is to be seen.

Overall, in line with recent IPCC reports, we have argued that taxes are to be preferred over the cap-and trade system as the latter is subject to many drawbacks such as their complicated institutional structure, the lack of its general implementability, and the volatility of the carbon price. As we have shown, in particular, given the recent stock market meltdown and volatility, the cap-and trade system presumably will not be very effective in providing sufficient incentives for reducing CO_2 emissions.

References

Brock, W. A.; Taylor, M. Scott (2004): Economic Growth and the Environment: A Review of Theory and Empirics. NBER Working Paper, 10854.

Coase, R. (1960): The Problem of Social Cost. Journal of Law and Economics, vol. 3: pp. 1-44.

Deissenberg, C.; Dawid, H.; Sevcik, P. (2006): Gullibility and Welfare in an Environmental Taxation Game. Computing in Economics and Finance 505.

Foley, D. K. (2008): The economic fundamentals of global warming. In: Harris, J.M.; Goodwin, N.R. (eds.): Twenty-First Century Macroeconomics: Responding to the Climate Challenge. Cheltenham/Northampton: Edward Elgar Publishing.

Greiner, A.; Semmler, W. (2008): The Global Environment, Natural Resources and Economic Growth. Oxford: Oxford University Press.

Heal, G. (1998): Valuing the Future: Economic Theory and Sustainability. New York: Columbia University Press.

Huber, C.; Wirl, F. (2005): Voluntary Internalisations Facing the Threat of a Pollution Tax. Review of Economic Design, vol. 9: pp. 337-362.

IPCC (2006, 2007): Climate Change 2007. UNEP.

Lovins, A. B.; Lovins, L. H.; von Weizsäcker, E. U. (1997): Factor Four. Doubling Wealth — Halving Resource Use. A Report to the Club of Rome. London: Earthscan.

Lovelock, J. (2006): The Revenge of Gaia. New York: Basic Books.

Mankiw, N. G. (2007): One Answer to Global Warming: A New Tax. NY Times, 07/09/16.

Metcalf, G. E. (2007): A Proposal for a U.S. Carbon Tax Swap. Washington D.C.: Brookings Institutions Press.

McKibbin, W. J.; Wilcoxen, P. J. (2002): Climate Change Policy after Kyoto: A Blueprint for a Realistic Approach. Washington D.C.: Brookings Institution Press.

Nordhaus, W. (2008): A Question of Balance. Princeton: Princeton University Press.

Pigou, A. C. (1932): The Economics of Welfare. London: Macmillian.

Semmler, W. (2006): Asset Prices, Booms, and Recessions. Berlin: Springer.

Stern, N. (2007): The Economics of Climate Change: The Stern Review. Cambridge: Cambridge University Press.

Uzawa, H. (2003): Economic Theory and Global Warming. Cambridge: Cambridge University Press.

Weitzman, M. (2008): On Modeling and Interpreting the Economics of Catastrophic Climate Change. working paper.

Greening the Economy in the European Union

Achim Brunnengräber

Introduction

In order to harmonise its climate and energy policies, the European Commission (EC) has set itself ambitious goals which do not lack determination. In order to counteract the political and economic risks involved in the dependence of the European Union (EU) on energy imports, the EC wishes to start a new industrial revolution. Low-carbon growth is to be accelerated, own energy production dramatically increased and competitiveness maximised. This is to be achieved by increasing the share of renewable energies in total energy consumption by 20% and reducing CO_2 emissions by 20%, both by 2020. The EU wishes to be a pioneer and model for sustainable development in the 21st century. Europe's agreement on precise, legally binding targets symbolises its determination to reconcile climate policy and energy policy. The Commission wishes to shape the Europe of 2050. In the view of the EU, the success of this enterprise will depend not least on how respectfully we treat the world around us. This is set out in the strategy papers of the European Union (European Commission 2007: An Energy Policy for Europe; European Commission 2008a: 20 20 by 2020. Europe's climate change opportunity[1]).

In all three areas of the energy and climate policy triangle of objectives – environmental friendliness, competitiveness and supply security – the Treaty of Lisbon has made considerable adjustments to the European Union's range of potential action and opportunities for influence (Treaty of Lisbon, particularly Art. 191 and 194; cf. also Fischer 2009). With the formula „20 20 by 2020", which was first agreed at the Spring summit of the 27 Heads of State and Government in 2007, the objective has finally been quantified. The conflicting targets in the energy strategy are not mentioned. However, a more detailed analysis of the energy strategy documents rapidly brings to light the politico-economic and ecological contradictions of a policy which attempts combining climate issues with energy security. The EU cannot plausibly explain how it is going to

---

[1] References in the following in which there are no particulars as to the literature refer to these two publications by the European Commission.

stake its claim to being a pioneer in climate protection while at the same time improving the supply situation with regard to fossil fuels and increasing the admixture of biofuels. In order to become a competitive *global player* in line with the Lisbon Strategy, considerable amounts of both fossil fuels and renewable resources will be required. A large share of these cannot be produced in the internal market, however, but will have to be imported. The overlapping of *internal* and *external* dimensions in its energy and climate policies thus turns out to be a characteristic structuring principle of the European Union, but certainly not a harmonised one.

If we leave the world of pronouncements and turn to the basic politico-economic position of the European Union that characterises the EU strategy papers, three interdependent maxims attract our attention: 1) the logic of competitiveness, 2) the belief in technological solutions, and 3) the exploitation of strategic resources of the developing countries. These maxims are not new. They have long determined the common environmental, trade and foreign policy of the European Union. No explanation is offered, however, as to how a successful climate protection and energy policy is to be combined with a successful growth strategy and competition policy. Instead, the field of politics remains opaque (Geden 2008: 354). In the following it will be argued that the question of the coherency of the different spheres cannot be solved by the strategies and instruments presented by the European Commission in its climate policy documents.[2]

1   Determination on Paper

At the G8 talks in June 2007 in Heiligendamm and at the climate negotiations in Bali in December 2007 the EU announced that it would reduce its emissions of greenhouse gases by 2020 by 25–40% compared to 1990. Following internal arguments it was forced to correct its reduction target downwards to 20% by 2020; this was combined with the announcement that the target would be increased to 30% if other industrial countries agreed to participate in the ambitious reduction target. Over the same period the share of renewable energies in final energy consumption in the EU was to be increased from 8.5% (2005) to 20% (2020). In June 2010 the EU Commission retreated from this ambitious target following internal criticism and massive criticism by industry. According to the Commission a CO_2 reduction of 30% by 2020 would mean a reduction in eco-

---

[2] Whereby the urgency of the challenge is without question: the NASA speaks of a timeframe of 10 years in which an effective climate protection policy would have to be effective in order to prevent only the greatest catastrophes. The Intergovernmental Panel on Climate Change (IPCC) points in particular to the ecological and socio-ecological effects of climate change (IPCC 2007).

nomic output in the European Union by 0.54% of gross domestic product. That would mean 81 billion euro – eleven billion more than the achievement of the 20% target would cost.[3]

Already in the EU Strategy of 2008 it is unmistakeably formulated that the central aim is the safeguarding of the „welfare of the European economy" (European Commission 2008a). This is necessary because the growing competition for access to fossil fuels, the worldwide increase in the price of oil, coal and gas, and economic losses due to the effects of climate change are causing considerable political conflicts in the world. As its answer to this the EU Commission wishes to lead the way with the creation of a low-carbon economy with high energy efficiency. At the same time, however, this objective is relativised: fossil fuels will remain the most important source of energy worldwide in the decades to come. The strategy papers state that it will remain necessary to have recourse to coal deposits, and even nuclear energy will contribute to the energy mixture – at the discretion of the Member States. Phrases such as these do not announce a radical transformation of the energy systems, nor do they proclaim a climate-friendly economic system. Neither the question of the finite nature e.g. of *peak oil*, nor the geostrategical conflicts involved in fossil fuels, nor the consequences of such a policy for the global increase in CO_2 emissions and thus the intensification of climate change are analysed. It is not determined whether the increasing requirements of fossil fuels in the EU can at least be compensated by the expansion of renewable energies. In 2005 the Member States imported 57% of their natural gas requirements and 82% of their oil requirements – for 2030 the EC even reckons with 84 and 93% respectively. Under a *business-as-usual* policy the dependence of the EU on carbon energy imports will therefore increase in total (European Commission 2007: 3).

What appears to be more realistic in the EU strategy is that the expansion of renewable energies and the necessity of reducing dependence on imports of fossil fuels can be used as a catalyst for the modernisation of the European economy. High energy prices will lead to the development of new and more efficient technologies. Reacting early to this situation will not only bring *first mover advantages*, but it will also secure a global competitive advantage in the long term (see also European Commission: Global Europe. Competing in the world, 2007). Thus, the aim of the EU's pioneering policy is not to achieve a balance in the unequal global economic relationships between North and South, to dismantle climate-damaging trade relationships or the existing global energy chains, nor is it to reduce the socially unequally distributed effects of ecological conflicts, but

---

[3] „EU weighs pros and cons of tougher emissions targets", European Commission, http://ec.europa.eu/news/environment/100527_en.htm (05/27/2010).

primarily to establish the position of the EU vis-à-vis its global competitors (cf. also Altvater/Mahnkopf 2007: 188ff).

The annual G8 summits also go along with this line of argumentation. In 2005 in Gleneagles the then UK Prime Minister Tony Blair described climate change as „probably long-term the single most important issue we face as a global community". He called for a struggle against climate change as the top-most priority (*„Tackling Global Climate Change"*). As a result, the energy and environmental ministers of the G8 countries then laid the foundation for a plan of action on climate change, clean energy and sustainable development which was adopted by the Heads of State and Government at the 2005 summit. The action plan also contains general declarations on climate protection and the urgency of action. On the subjects of energy efficiency and energy saving numerous sup-portive measures were listed, but neither the required quantitative targets and timeframes nor legislative initiatives or specific measures for the renewable energies' development are named explicitly.

The Gleneagles action plan was anything but a clear commitment to the rad-ical transformation of the energy systems. It foresees neither the abandonment of nuclear energy nor the abandonment of energy from coal and other fossil fuels. Instead, the G8 would „support efforts aimed at making the production of elec-tricity from coal and other fossil fuels cleaner and more efficient". Only one year later in St. Petersburg, however, the development and strengthening of climate protection measures had already been displaced by questions of global energy security.[4] In the energy action plan presented by the G8 countries climate protec-tion no longer played a major role. Instead, thousands of billions of dollars of investment in energy security in the following 25 years and the massive expan-sion of oil and coal production were regarded as necessary. The International Energy Agency (IEA) calculated that 12.5 million additional barrels per day were needed to cover world requirements. For this reason alone, oil production projects, particularly in the key countries of the OPEC, would be necessary (cf. Birol 2008).[5] Although the issue of climate protection was again put at the top of the political agenda at the G8 summit in Heiligendamm in 2007, it once more proved impossible to agree on a common strategy on the sustainable use of re-sources, on the development of renewable energies or on climate protection which included concrete, binding and verifiable measures and targets. Climate protection takes second place to energy policy (Brunnengräber 2007).

In contrast, the primary orientation of energy policy to supply security has a long tradition. As early as 1975 in Rambouillet the Heads of State of the then 6

---

[4] Announcement on the official website of the G8 summit 2006: http://en.g8russia.ru/agenda (down-load: 02/10/2007)

[5] http://www.energiestiftung.ch/newsletter_nr_12.htm (download: 14.2.2007)

most economic powers in the world (G6) felt obliged to discuss the controversial subject of „ energy security" informally at the highest level. Already at that time the following clear statement was made: „World economic growth is clearly linked to the increasing availability of energy sources. We are determined to secure for our economies the energy sources needed for their growth. Our common interests require that we continue to cooperate in order to reduce our dependence on imported energy through conservation and the development of alternative sources."[6] At that time the first oil price crisis was decisive for this canvassing for energy saving, energy efficiency and renewable energies.

Today it is climate protection which apparently forces action. The powerful industrial countries, the OECD, the G8 and the European Commission also indicate, however, that a stable supply of oil and gas must be secured, that for this purpose new supply regions must be found (import diversification) and investments made in new and improved pipelines and storage sites. At the same time reports are increasing that the gap between global energy hunger and energy scarcity will continue to grow: supply bottlenecks are threatening, prices are escalating and inflation is increasing. The EU's strong fixation on low energy prices and its adherence to fossil fuels as a precondition for economic stability and growth are thus based on a fragile foundation. But how should the now obvious contradictions between climate and energy policy be dealt with, and how should the potential conflicts of objectives be kept within bounds?

2 Separate Energy Regimes

Energy and climate policies have developed differently within the political process and have until now remained institutionally apart. With regard to climate change, the focus is on CO_2 emissions, the effects of which are global. However, these effects concern (only) the *output* side of the fossil energy system. As a result, in the dominant discourse climate change is described, defined and politically construed as a global environmental problem. If, in contrast, the focus lay on the consumption of the energy sources coal, gas and oil, i.e. on the *input* side of fossilism, then the regional and national interests would also have to be taken into account (Brunnengräber 2006). The dominant interpretation of the problem is decisive for the separation of climate policy not only from European energy policy, but also from trade and finance policy as developed by the World Trade Organisation (WTO) and the International Monetary Fund (IMF). Thus, the pov-

---

[6] Declaration of Rambouillet, 11/17/1975, http://www.g8.fr/evian/english/navigation/the_g8/previous _g8_summits_in_france/rambouillet_-_1975.html#com (download: 06/21/2010).

erty policy of the World Bank continues to have its foundation in a growth paradigm based on fossilism and modernisation theory. Support for renewable energies by the World Bank Group is only 5% of its total investment in the energy sector. Expenditures on projects in the field of fossil fuels, in contrast, were generously increased by 93% from 2005 to 2006 (from 450.8 to 869 bn. US$) (Setton et al. 2008). The climate policy funds for adjustment support at the beginning of 2010 just amounted to US$ 350 bn.

The political, economic and social institutions of the fossilistic-capitalist economic system, the policies of which are oriented towards growth and competitiveness, disappear from view as perpetrators of climate change due to this institutional separation. The social inequalities between those who have high luxury emissions and those who only emit as much as they need to survive are also hardly recognisable from the global perspective. If the different dimensions of the problem are taken into account, then climate change is not only an ecological problem but a representation of the comprehensive geostrategic, social, political and economic crisis of society (Brunnengräber et al. 2008, Brunnengräber 2011). The institutional separation is also reflected, however, in the European Union and the Lisbon Treaty: until now the European Union has only had a very weak portfolio of primary rights with regard to energy policy, which is only „incompletely communitised" (Altvater/Mahnkopf 2007: 196), while climate policy, on the other hand, is becoming increasingly „Europeanised" (for example in the development of renewable energies or in emissions trade). This policy of „two speeds" can clearly be seen in an international context, too.

The UN Framework Convention on Climate Change (1992) and the Kyoto Protocol (1997) created an international policy field, the design of which only influences national energy policy indirectly via emissions trading. A paradoxical situation has arisen here in the area of renewable energies. While their development – even so this is absurd from the point of view of climate policy – does not play a major role in international climate policy (they produce only few emissions), their strategic importance at a national and European level is growing. Because they are available nationally, dependence on imports of fossil energy sources can be reduced by their use and supply security thus improved. In this order, first energy policy then climate policy, the European Union takes up this subject in its „White Paper: Energy for the future – renewable sources of energy" (European Commission 1997). However, the sovereignty of the Member States over their primary sources of energy and over their energy (security) policies remains untouched. In particular the Central European states under the leadership of Poland are opposed to a too ambitious emissions trade because they wish to continue using their high-emission mineral coal. The G8 countries are also far from achieving uniform or coherent energy and climate policies.

Finally, the lack of coherence between energy and climate policy can also be seen in the development of emissions. The EU, with its share of 14% of global energy consumption and the ten tonnes of CO_2 which each EU citizen produced on average in 2007, finds itself in mid-field with regard to climate policy. In 2009 the emissions of the EU-15 fell by 5,8, compared to the previous year (Ziesing 2010), but in the Member States the emission of greenhouse gases are very different. Whereas the UK (-15.8%), Germany (-20.8%) and France (-11.8%) were able to reduce their CO_2 emissions, they increased markedly in Spain (+55.3%), Portugal (+30.8%), Austria (+ 7.6%) and Italy (+ 7.41%) (www.unfccc.de). Almost all the emission data thus indicate the increasing emission of greenhouse gases, so that it appears improbable that the reduction targets, which are binding in international law, can still be achieved in absolute figures by 2012 (DIW-Wochenbericht 35/2006, Ziesing 2010). The European Commission, like the G8, therefore speaks less and less often of the necessary reduction of CO_2 emissions but of the fact that all states must make a contribution towards limiting the increase in global average temperature to less than two degrees Celsius above the pre-industrial level.

3 The Control of Nature through New Technologies and New Markets

Despite this fact, the EU Commission's report of 11/12/2009 „Progress towards achieving the Kyoto objectives" (COM(2009)630) claims that the Kyoto target of 8% by 2012 will be achieved. This will only be the case, however, as the Report explains, if the storage capacity of forests, the market economy instruments (the EU emissions trade and the Clean Development Mechanism) and the emissions trade between countries as intended by the Kyoto protocol are put into practice. The exploitation of the loopholes will lead to target achievement in the sense of creative carbon bookkeeping, but not necessarily to an absolute reduction in CO_2 emissions. We therefore repeat the question we put at the beginning: how does the EU intend to translate its determination into deeds, to realise its ambitious reduction targets and to increase competitiveness?

Few measures for the reconciliation of climate and energy policy which go beyond the Kyoto protocol can be found in the strategy papers. There is one reference to the obligation to label energy-saving electrical goods, and many other references to the importance of energy efficiency. The technology of carbon capture and storage (CCS) is dealt with much more extensively. It is not mentioned, however, that there are considerable uncertainties in this connection and that this technology will not be available for ten to fifteen years. The considerable energy requirements of this technology and the potential risks that arise when the greenhouse gases are pressed into the Earth's waterfalls are not discussed, although the protest from affected members of the public against the pilot projects increase. There is a reason for this: the „20 20 by 2020" strategy of the European Commission stands on shaky ground without the storage of CO_2, as fossil-fired coal power stations are intended to play an even more important role in the production of energy in the future. Furthermore, two birds are killed with one stone: domestic stocks of coal can in this way be used within the EU in a „climate-friendly" fashion, and the new technology can be sold to developing countries. The EU is presently supporting the establishment of up to 12 demonstration plants by 2015. It is not taken into consideration here, however, that each investment in the new technologies will mean the prolonging of the fossil energy regime for decades and in the long term will even hinder a radical transformation into a new energy age.

In addition to the CCS technology, the strategy paper emphasises the „positive experiences" with the trade in emissions and with the Clean Development Mechanism (CDM). For the EU the Kyoto protocol represents above all a set of political regulations which creates the framework for new, lucrative markets (Brunnengräber 2006). This orientation of international climate policy is already anchored in the UN Framework Convention on Climate Change: „The parties should cooperate to promote a supportive and open international economic system that would lead to sustainable economic growth and development for all Parties, particularly developing country parties (…)." (UNFCCC Artikel 3.5). According to the European Commissioner for the Environment, Stavros Dimas, emissions trade shows „our global partners that strong action to fight climate change is compatible with continued economic growth and prosperity. It gives Europe a head start in the race to create a low-carbon global economy that will unleash a wave of innovation and create new jobs in clean technologies."[7] The problems connected with emissions trade, such as the surplus and the low price of the certificates, are no longer discussed (Schüle 2008) and the extensive criticism to which CDM has meanwhile been subjected continues to be ignored (Hallström et al. 2006). The CDM is directly connected with emissions trade via the *EU Linking Directive*, which enables European businesses to cover up to 25% of their reduction obligations with certificates which have been generated by investments in, and the development of, renewable energies in the developing countries. The EU intends to increase the upper limit to 30%.

---

[7] „Climate change package to enhance growth & jobs" Policy News, European Commission, http://ec.europa.eu/environment/etap/inaction/policynews/163_en.html (06/17/2010)

De facto the CDM has shown in recent years, however, that the flexible mechanism is scarcely able to fulfil the expectations placed in it with regard to CO_2 reductions and the promotion of sustainable development. At best the CDM can be described as a zero-sum game, because higher emissions in the North are balanced out by emission reductions in the South. However, even that is questionable. More than a third of the tradable certificates are generated by *end-of-pipe* technologies. Particularly problematical is the generation in HFC-23 projects: the gas which results from the production of refrigerants has a high potential for „global warming" and is therefore an extreme climate-killer. A large number of emission certificates can be generated quickly and cheaply by the combustion of this gas.[8] This awakens greed: there is a danger that the Kyoto instrument will create an economic stimulus to produce climate-killers in order to dispose of them – at a profit – at a later point in time.

The geographical distribution of the CDM measures is also an indication of the discrepancy between expectation and reality of the mechanism. More than 90% of the tradable certificates in the sphere of CDM come from India, China, South Korea and Brazil. In most of the rural regions of Africa, Latin America and Asia there are only few CDM investments. Here too, the sustainable transformation of energy systems or the reduction of the worldwide „energy poverty" by the development of decentralised renewable energy supply systems does not take place in the CDM. Investments in renewable energies flow primarily, because the transaction costs are high, into large-scale projects such as dams. These and other CDM measures such as afforestation in the form of monocultures or the support for genetically altered varieties of rice which require less climate-damaging nitrogen fertiliser, also produce new local and regional conflicts not recorded in the carbon bookkeeping of the North.

4   Strategic Raw Materials Policy: Agrofuels

Agrofuels are also conflict-laden. For the European Commission they represent „the only realistic alternative fuel in the foreseeable future" which combines environmental sustainability with firm growth criteria in a climate-friendly way. Here too, it is clear that the EC is consolidating fossilistic energy supply rather than questioning it. The transition to agrofuels causes a number of problems (cf. also WBGU 2009): *firstly,* the admixture prolongs the use of fossil fuels, which

---

[8] The „global warming" potential of HFC-23 is 11,700. That means that for each tonne of HFC-23 which is not emitted 11,700 certificates are generated (for more details on this problem, see Pearson 2007).

can hardly be described as a climate-friendly strategy. *Secondly*, competition arises between „food, feed and fuel", between food for hungry mouths, feed for animals and fuel for empty tanks, as the same sources of energy, maize and soy, sugar cane and palm oil, rape seed and turnips serve as food for humans and are transformed in agriculture into meat products and into fuels for automobiles. *Thirdly*, the reduction of CO_2 emissions in the transport sector via the admixture of biofuels has proved to be more or less a zero-sum game. If the complete life cycle of the fuels from planting through production to transport is taken into account, the result is often no better than in the case of conventional fossil fuels (IFPRI 2006, IFPRI 2010).

The worldwide price increases of basic foodstuffs have already pointed out dramatically the real and future conflicts between nutrition security and sovereignty, especially in the South, and energy security in the North. The IFPRI calculates that the prices of basic foodstuffs, which are of great importance above all in African, Latin American and Asian countries, will continue to show an upward trend until 2020 due to increasing demand: maize by c. 40% and manioc by 135% (IFPRI 2006 in: Fritz 2007). At the Spring Conference of the World Bank in Washington in 2008 the director, Robert Zoellick, declared that about 100 million people in the developing countries could slide further into „destitution" because of the worldwide price increases of basic foodstuffs while 33 countries could be afflicted by social chaos and political unrest.[9] Haiti, Indonesia, Mexico and Bangladesh would then only be the forerunners of a development revealing the destructive power of competition between foodstuffs for humans, animals and motors.

Against this background the objective of the EU to achieve a share of admixture in the transport sector of 10% by 2020 (EU-Directive 2009/28/EC) is two-faced: on the one hand this is intended to contribute to climate protection, but on the other hand it leads to a quite deplorable state of affairs: to conflicts over land use, the deforestation of rainforests, price increases for basic foodstuffs, the loss of the variety of species, the expansion of monocultures and the intensive use of pesticides. In future this should be reacted to by binding criteria for the preservation of the variety of species and certain forms of land use, but doubts as to the enforceability of these criteria are justified. Examples such as the Forest Stewardship Council show that transnational concerns massively violate the basic principles of this charter – even though they nowadays integrate sustainability criteria and environmental and social standards into their corporate concepts.

---

[9] „Food Price Surge Could Mean ‚7 Lost Years' in Poverty Fight"; press statement by World Bank President Robert B. Zoellick (www.worldbank.org, download: 04/14/2008).

The EU already imports up to 90% of its vegetable raw material require-ments from third countries. It is not the only power with such a demand which is interested in the agrarian energy raw materials of the developing countries, how-ever. At the same time worldwide competition for arable land is increasing, land which is also becoming ever more important for the supply security of China, the USA and the oil countries of the Middle East. In the final analysis, if binding socio-ecological standards are dispensed with, resource requirements and the world market will decide whether the cultivation of biomass such as maize, palm oil or soy is worthwhile – and not their effects on climate protection. Already a global politico-economic energy complex has arisen consisting of finite energy sources, the agro-business for renewable raw materials, a market for gene tech-nology products and renewable energy sources, which is attempting to guarantee energy supply and thus growth, despite all the apocalypse scenarios with regard to oil. And it is clear how the economic profits and the high living standard of the North are to be secured: by the sites for the production and extraction of raw materials in the global South.

5   Overcoming the Oil Age

Since the beginning of industrialisation economic growth, locational competition and economic stability have been based on the availability of fossil energy re-sources. There is a political congruence between fossil energy systems and the globally dominating capitalist economic system (Altvater 2006). But congruence is lacking in the sense of a solution-oriented climate protection policy. Instead, their incompatibility is secured politically, geostrategically and, if necessary, militarily. The „firewall between the economic and the ecological energy re-gime" (Altvater 2005: 82) is therefore established in a political process and is not at all a natural necessity. If friction arises, there is a reaction. The strategy paper „Climate Change and International Security" drawn up by the EU High Repre-sentative for Common Foreign and Security Policy, Javier Solana, at the request of the European Parliament should be interpreted in this sense. In it climate change is understood as a „threat multiplier" which „exacerbates existing trends, tensions and instability". In a world of growing threats to security the EU should prepare itself for competition over energy sources and control over these, for an „increased migration pressure" from South to North, and for struggles over the distribution of land which are largely caused by climate change.[10] By means of a preventive policy the costs or „negative" consequences of climate change for the EU should be avoided, trade relationships expanded and the sources of raw materials secured in the long term (European Commission 2008b). By means of such a security discourse climate change is transformed from a soft issue into a „hard" field of politics (cf. also Geden 2009: 9).

Socio-ecological and energy-policy necessities are subordinated to the objectives of strategic, economic and, henceforth, security policy rationality. Consequently, the internationally agreed mechanisms of the Kyoto protocol will not permit „a reduction of emissions which is greater than the rhythm of economic growth allows" (Leff 2002: 102; cf. also Hallström et al. 2006). Both the EU and the G8 strategies are designed for economic stability, economic growth and welfare in the Member States. This is to be achieved by technological innovations in the sphere of carbon storage and fuel development on the one hand and the expansion of flexible market mechanisms in the sphere of emission reductions on the other. In this way, the EU Commission wishes to lay the foundations for a climate capitalism without having to give up the advantages of a fossil capitalist social system. The Lisbon Treaty maintains integration in the sphere of environmental policy but leaves the security of energy supply in the hands of national decision-makers. So far, the Commission has „flinched from the application of concrete legislative procedures" here (Fischer 2009: 57f).

The strategy papers, too, tend to express the energy policy continuities in national energy policies rather than „revolutions". The greatest potential for supply security is to be delivered by an energy mixture in which nuclear energy is given a concurrent role alongside coal, gas and oil. Nuclear energy is, furthermore, regarded as important for the achievement of a significant reduction in CO_2 emissions. This is a reflection of the EU's strong fixation on low energy prices as a condition for economic stability, on economic growth and the realisation of the Lisbon strategy, namely to become the most competitive and most dynamic economic area in the world. However, the EU cannot achieve the increase in its competitiveness under its own steam. It does not have enough natural resources of its own.

A different perspective should therefore be adopted: „It is high time, for the purposes of debate and policy-making, to put the spotlight on the core-problem – fossil fuel extraction and consumption" (Hallström et al. 2006: 2). If climate protection and the reduction of CO_2 emissions are seriously formulated as targets, then the oil age must be overcome; and this must happen before the scarcity

---

[10] As early as 2003 a Pentagon Report discussed the potential effect and intensity which the results of climate change could have on national security issues via environmental refugees (Schwartz/Randall 2003).

of fossil energy sources ensues. The fundamental importance accorded to fossil energies, energy security and low energy prices hardly points in this direction, however. Neither in international climate policy nor in the EU is there any sign of the political will to conduct a radical transformation of the energy systems towards renewable and decentralised energy sources and thus away from the dominant fossilistic, centralised energy supply structures.

References

Altvater, E. (2005): Das Ende des Kapitalismus wie wir ihn kennen. Münster: West-fälisches Dampfboot.

Altvater, E. (2006): The Social and Natural Environment of Fossil Capitalism. In: Panitch, Leo; Leys, Colin (eds.): Socialist Register 2007: Coming to Terms with Nature. London/New York/Halifax: The Merlin Press, pp. 37-71.

Altvater, E.; Mahnkopf, B. (2007): Konkurrenz für das Empire. Die Zukunft der Europäischen Union in der globalisierten Welt. Münster: Westfälisches Dampfboot.

Birol, F. (2008): Interview with Astrid Schneider under the title „Die Sirenen schrillen". Internationale Politik, April 2008: pp. 34-45.

Brunnengräber, A. (2006): The political economy of the Kyoto protocol. In: Panitch, Leo; Leys, Colin (eds.): Socialist Register 2007: Coming to Terms with Nature. London/New York/Halifax: The Merlin Press, pp. 213-230.

Brunnengräber, A. (2007): Energiesicherheit vor Klimaschutz. In: Melber, Henning; Wilß, Cornelia (eds.): G8 Macht Politik: Wie die Welt beherrscht wird. Frankfurt am Main: Brandes & Apsel, pp. 113-123.

Brunnengräber, A. (2011): Multi Level Climate Governance. In: Knieling, Joerg; Leal Filho, Walter (eds.): Climate Change Governance. Frankfurt: Springer, forthcoming.

Brunnengräber, A.; Dietz, K.; Hirschl, B.; Walk, H.; Weber, M. (2008): Das Klima neu denken. Eine sozial-ökologische Perspektive auf die lokale, nationale und internationale Klimapolitik. Münster: Westfälisches Dampfboot.

European Commission (1997): White Paper: Energy for the future – renewable sources of energy. Brussels: Communication from the Commission, COM(97) 599 final.

European Commission (2007): An Energy Policy for Europe. Brussels: Communication from the Commission to the European Council and the European Parliament, COM (2007) 1 final.

European Commission (2008a): 20 20 by 2020. Europe's climate change opportunity. Brussels: Communication from the Commission to the European Parliament, the Council, the European Economic and Social Committee and the Committee of the Regions, COM(2008)30 final.

European Commission (2008b): Climate Change and International Security. Brussels: Paper from the High Representative and the European Commission to the European Council, S113/08.

Fischer, S. (2009): Energie- und Klimapolitik im Vertrag von Lissabon: Legitimierungs-erweiterung für wachsende Herausforderungen. Integration, No 01/09: pp. 50-62.

Fritz, T. (2007): Das Grüne Gold. Welthandel mit Bioenergie – Märkte, Macht und Monopole. Berlin: FDCL.

Geden, O. (2008): Die Energie- und Klimapolitik der EU – zwischen Implementierung und strategischer Neuorientierung. Integration No 04/08: pp. 352-364.

Geden, O. (2009): Klimasicherheit als Politikansatz der Europäischen Union. Dskussionspapier der FG 1 EU-Integration, Stiftung Wissenschaft und Politik, 2009/01, SWP Berlin.

Hallström, N.; Nordberg, O.; Österbergh, R. (2006): Carbon Trading. A critical conversation on climate change, privatisation and power. Uddevalla: Mediaprint.

IFPRI (2006): Bioenergy and Agriculture: Promises and Challenges. International Food Policy Research Institute. 2020 Focus No 14, November 2006, Washington D.C.

IFPRI (2010): Global Trade and Environmental Impact Study of the EU Biofuels Mandate. Washington D.C.: International Food Policy Research Institute.

IPCC (2007): Climate Change 2007: Climate Change Impacts, Adaptation and Vulnerability. WG II Contribution to the IPCC Forum Assessment Report. Summary for Policymakers. Available at www.ipcc-wg2.org/index.html, accessed April, 10th 2007.

Leff, E. (2002): Die Geopolitik nachhaltiger Entwicklung. Ökonomisierung des Klimas, Rationalisierung der Umwelt und die gesellschaftliche Wiederaneignung der Natur. In: Görg, Christoph; Brand, Ulrich (eds.): Mythen globalen Umweltmanagments: "Rio + 10" und die Sackgassen nachhaltiger Entwicklung. Münster: Westfälisches Dampfboot, pp. 92-117.

Pearson, B. (2007): Market failure: why the Clean Development Mechanism won't promote clean development. Journal of Cleaner Production, 15 (2007): pp. 247-252.

Schüle, R. (ed.) (2008): Grenzenlos Handeln? Emissionsmärkte in der Klima- und Energiepolitik. Munich: oekom.

Schwartz, P.; Randall, D. (2003): An Abrupt Climate Change Scenario and its Implications for United States National Security. Available at http://www.gbn.com/GBNDocumentDisplayServlet.srv?aid=26231&url=%2FUploadDocumentDisplayServlet.srv%3Fid%3D2 8566, accessed December, 10th 2003.

Setton, D.; Knirsch, J.; Mittler, D.; Passadakis, A. (2008): WTO – IWF – Weltbank. Die "Unheilige Dreifaltigkeit" in der Krise. Hamburg: VSA-Verlag.

WBGU (2009): Future Bioenergy and Sustainable Land Use. German Advisory Council on Global Change (Wissenschaftlicher Beirat der Bundesregierung Globale Umweltveränderungen), Berlin.

Ziesing, H.-J. (2010): Wirtschaftskrise beschert Rückgang der weltweiten CO2-Emissionen. Energiewirtschaftliche Tagesfragen, No 09/2010.

The „Tragedy of the Atmosphere" or the Doubling of the Carbon Cycle and the Circulation of Capital

Elmar Altvater

Introduction

Ever since the research conducted by the Swedish physicist and chemist Svante August Arrhenius towards the end of the 19th century it has been known that the increase in temperature is connected with the increase in the concentration of CO_2 in the atmosphere. This makes it possible, according to Arrhenius, for humans „to live under a warmer sky". His optimism appears to us today to be naive We now know that the emissions of carbon dioxide together with other greenhouse gases (GHG) into the atmosphere and the resulting increase in the average temperature of the Earth has some very unpleasant results, it even can trigger catastrophic consequences for the evolution of life on Earth, for the peaceful reproduction of human societies, for economic wealth and welfare. The climate-conference of Cancún in December 2010 agreed on an obligatory limitation of an increase of the planet's average temperature of 2° Celsius above the pre-industrial level. In order to reach this objective it is necessary to reduce the increase of greenhouse gas-emissions and to respect the level of 450 *parts per million air molecules* (ppm). That will only be possible, and we know this too, if fossil fuels are no longer burned as they have been since the fossil-industrial revolution in the second half of the 18th century.

If the deposits of hydrocarbons were depleted before the increase in the concentration of their combustion residues in the atmosphere led to an increase in the average temperature of the Earth to the point at which the planetary climate system collapses, then we would have solved the energy and climate crisis at one blow. Unfortunately, however, it is very likely that Günther Anders is right: the Earth is like an exploitable mine, and therefore it must be exploited (Anders 1995: 32). The fossil fuels will be extracted from the Earth in their entirety – and remain, after they have been burned, as greenhouse gases in the planet's atmosphere. It would be economically irrational not to do this. Fossil fuels suit the capitalist mode of production perfectly and give it the unprecedented dynamism of the two centuries since the industrial revolution. They have contributed to deeply changing the planet.

The International Energy Agency (IEA) assumes that the consumption of primary energy will increase by 55% by the year 2030 (an average annual rate of growth of 1.8%) (IEA 2007:3). In the OECD the figure will be 1.5% per annum according to IEA forecasts, and 2.3% in the non-OECD countries (IEA 2009: 76). The reserves of fossil fuels should be large enough to cover this demand until 2030, as long as the oil industry is prepared to invest more than US$ 20 trillion in the discovery, extraction, refining and transport logistics of fossil fuels (particularly oil). Investments of this size would be able to create a high-yield terrain for the speculators who were shaken up in the global financial crisis.

With regard to climate policy, an – even restrained – continuation of the fossil system would mean that the emissions of greenhouse gases would not be reduced in the course of the century to the extent needed to stabilise the climate. The IEA (2009: 195ff.) reckons that a stabilisation of the concentration of green-house gases in the atmosphere at a value of 350 ppm to 550 ppm is possible. It assumes a target of an average of 450 ppm, which is regarded by many climate researchers as much too high. This value can only be attained, however, if the consumption of fossil fuels, particularly oil, shows only a small increase. Oil consumption should increase by 0.2% per annum, reaching 89 mb/d by 2030. Additional fuel consumption up to 2030 should above all be provided by „zero-carbon fuels", i.e. by renewable energy sources, the share of which in global energy consumption should increase from 19% to 32% by 2030. At the same time on the emission side, CO_2 emissions in the OECD should cost 50 USD per tonne and in the non-OECD countries 30 USD per tonne (IEA 2009: 195; 208), in order to approach the stabilisation goal of 450 ppm from two sides: from the input of fossil energy and from the output of emissions. Whether in this way the reduction of greenhouse gas emissions regarded as necessary by the Intergov-ernmental Panel on Climate Change (IPCC) of 50% by 2050 will be achieved is uncertain, especially if both sides are in the final analysis left to „market-based instruments".

In the international climate debate little thought is given to the fact that the necessary reduction of greenhouse gas emissions will affect the global circula-tion of capital. If at the beginning of the energy chain reserves of fossil fuels are left unused in the ground this means that natural resources are not transformed into value (valorised) although this would be possible. If they have already been valorised and included as assets in the balance-sheets of firms then they will have to be removed from them, they have to be written off. This is equivalent to the annihilation of capital. The certificates which are issued at the end of the energy chain to the energy consumers and which grant them emission rights for greenhouse gases are also a transferable economic good which increases the capital assets of a firm and thus may increase its value on the stock exchange. The emissions certificate can be traded as a bond on specialised stock exchanges. The emission trade thus is following the same business model as that which failed in the great financial crash of 2008: originate tradable assets (emission certificates) and then distribute them on global markets. Since there is neither labour nor real capital necessary to originate these fictitious values, there is no limit of their increase.

In the following several aspects of doubling the fossil energy chain into a carbon cycle and into a valorisation cycle will be examined.

1 The Fossil Energy Chain – a Carbon Cycle and a Valorisation Cycle

The transformation of useful, carbon-bound energy into greenhouse gases can physically be interpreted as an irreversible increase in entropy through the transformation of materials and energy. The useful fossil energy (hydrocarbon) which is suitable for the production of use values is dissipated in the form of carbondioxid (CO_2) following combustion and cannot be used for this purpose a second time.

The emission of greenhouse gases takes place as an unintended result of decisions taken „locally" which are rational individually and from the point of view of private enterprise: the fossil fuels are used as input in the production process. This takes place completely in harmony with the rules of a capitalist market economy, i.e. it follows capitalist rationality. But the result, namely the transformation of the Earth into a greenhouse, represents a monstrous global irrationality. We are therefore dealing with a simply classical form of tragedy: the best intentions and a completely rational approach produce an undesired and because of its negative climate effects irrational result. The answer to the climate crisis at the end of the energy chain with the „market-based instruments" of emissions trade appears to be highly rational, but it represents a continuation of irrationality, just as does the policy of energy security which at the beginning of the energy chain promises to secure a supply of fossil energy which increases in correspondence to demand (on the institutional separation of energy and climate policy cf. Brunnengräber 2011). There can be no escape from the rationality trap using today's usual energy- and climate policy measures. The curtain is being raised for the „tragedy of the atmosphere".

One barrel of oil (159 litres or approx. 0.16 tonnes) is transformed by combustion by a factor of 2.5 into approx. 0.4 tonnes of CO_2. Currently more than 70 m barrels (more than 11.2 m tonnes) of oil (leaving coal and gas out of consideration) are transformed daily into approx. 28 m tonnes of CO_2. These emissions add up to more than 10.2 bn tonnes per annum which increase the quantity of CO_2 in the Earth's atmosphere. They remain in the atmosphere for about 120 years (Enquete Kommission „Schutz der Grünen Erde" 1994: 9). Human beings in different regions of the world are responsible to very different degrees for the emissions of CO_2 into the atmosphere. The USA currently produce 20 tonnes of CO_2 per capita per annum, while the EU averagely produces about 9 tonnes. In China the figure is currently 3.5 and in India approx. 1 tonne per capita. Measured in terms of greenhouse gas emissions, the „environmental footprint" is, firstly, too great to prevent a dangerous increase in temperature, and secondly, extremely uneven in today's world.

Carbon dioxide and the six other greenhouse gases explicitly listed in the Kyoto Protocol are responsible for the Earth's radiation budget. The quantity of CO_2 in the atmosphere measured in ppm has increased since the industrial revolution from approx. 280 ppm to more than 380 ppm. It should not increase to more than 450 ppm (according to the OECD 2008, the IEA 2009) in order to limit the increase in the average temperature of the Earth to 2°C. Some climate scientists, however, regard a reduction in the concentration of the CO_2 in the atmosphere to around 350 ppm as necessary; greenhouse gases which have already been deposited in the atmosphere would therefore have to be withdrawn. Whether that is possible, taking into account experiences with carbon capture and storage and with the protection of carbon sinks (protection of the forests within the framework of the REDD), may well be doubted.

At the beginning of the industrial revolution the natural deposits of fossil fuels contained approximately 5000 gigatonnes of carbon. That is not much compared to the more than 100 million gigatonnes to be found in sediments and oceans (cf. Kromp-Kolb/Formayer 2005: 133ff.). Nevertheless, their liberation via their application in the economy has had disastrous consequences for the planet's climate. Although CO_2 is partly „consumed" by the growth of plants, on the other hand it is also emitted by the dying of plants and by the combustion of biomass. CO_2 is partly absorbed by the oceans, either by vertical currents which „dump" the CO_2 into the deep sea, or by biological degrading. The regeneration mechanisms of the carbon cycle are uncertain (and in part also unknown). Once fossil fuels have been released from the closed reservoirs in the ground like a „spirit from a bottle" (Rahmstorf/Schellnhuber 2007: 133), the greater part of the CO_2 emissions remain in the atmosphere and is deposited there for a long duration: „once above ground, carbon constantly flows back and forth among vegetation, water, soils and air" (Lohmann 2006: 6; similarly Kromp-Kolb/Formayer

2005: 133ff). The policy of combatting the greenhouse gas effect is therefore not one for the short term. It requires stamina, duration and a planetary perspective.

The fossil fuels have many advantages for a capitalist system in comparison with solar, renewable energy (on this cf. Altvater 2005). Fossil energy is not only very compact, it also has – at least until the peak of output has been reached („peak oil", „peak coal" and „peak gas") – a high EROEI (energy return on energy invested). It is well suited for the acceleration of all processes and for the extension of the geographical reach of trade and for the continuous increase of the productivity of labour. Fossil energy fits the accumulation dynamism of capitalist societies. Hydrocarbon as a fuel therefore has a high use value. Transformed into CO_2 however, it is not only useless but damaging the climate and it is therefore noted in Appendix A of the Kyoto Protocol as being the most important of all six „greenhouse gases".

The transformation of useful energy chemically bound in carbon into greenhouse gases can therefore be interpreted physically as an irreversible increase in entropy in the transformation of materials and energy. Economics assume reversibility, while in nature all processes are directional and principally irreversible. The rhythms and periods of time are also different. The exploration of reserves of fossil energy and, even more so, their extraction, transport and transformation into consumable energy are the subject of profitability calculations by firms in the energy supply sector, which today are usually transnational corporations. Their time horizon is very much shorter than that of the formation of the fossil reserves (many millions of years) and the duration of CO_2 in the atmosphere. Above all, it is based on the interest rates formed on globalised financial markets. The higher these are, the more short-sighted do economic actors behave (myopism), because invested capital is redeemed in shorter periods. For this reason the (economic and political) factors influencing the level of interest rates are of an importance for energy and climate policy which should not be underestimated.

At the same time „market-based instruments" are being politically created, tradable on financial markets and stock exchanges by specialised actors. Economists, but also a large number of climate researchers (cf. e.g. the authors in Müller/Fuentes/Kohl 2007) believe there will be gains in efficiency in the production and consumption of energy and decisive progress in the reduction of CO_2 emissions through the trade in „oil futures" etc. and emission certificates. Whether this will take place, and whether it will in fact contribute to climate stability, is uncertain to say the least, if not impossible. It is certain, however, that emissions trade offers the participating private enterprises and a multitude of consultants and lobby groups an „opportunity to develop new areas of business", as the previous German Minister for the Environment of the grand coalition government,

Sigmar Gabriel, promised in the preface to a brochure of his ministry (BMU 2006). Climate policy is even by their agents understood as a contribution to the origination of financial assets.

All the elements and sections of the fossil energy chain are thus doubled. On the one hand the physical, chemical and biological irreversible transformations. Oil reserves in the Earth's crust are pumped to the surface through the borehole, then brought to the refineries through pipelines and with the aid of tankers, converted into useful energy, which is transferred with the aid of complex distribution networks to the tanks of cars etc., which not only transform the petrol into mechanical power for the purpose of motion but also emit CO_2 (and other gases) into the atmosphere. This transformation inevitably leads to there being less oil in the reserves and more CO_2 in the atmosphere. That is the cycle of „wet oil", which fulfils itself in „gaseous emissions" which then run through their own carbon cycle as greenhouse gases. Nobody would consider producing petrol and crude oil from the CO_2. Everybody knows that physical processes are irreversible, or they at least suspect so.

In the economy, however, fossil energy circulates as an economic value, which continually returns to itself, increased by profit. Profits are, as we say precisely in English, „returns to capital". The reversible capital cycle must therefore take the form of a spiral if it is to satisfy its own rationality criteria. The price of a barrel of oil which appears in the balance-sheet of a firm is tied to the physical oil, the „wet oil". The value of the oil also circulates independently in the form of „paper oil", however, and is traded on specialised forward and futures markets on the commodity exchanges in Chicago, New York and London. By looking at the money it cannot be seen, from the sales of which commodities it has come into the hands of the money owner. Paper oil is securitised titles, bonds concerning the ownership of the content of tankers, concerning claims to interest or dividend payments in the present or concerning sales and purchases of a certain commodity (in our case oil) in the future. Once they have been securitised, bonds for oil shipments or oil reserves and rights to the pollution of the atmosphere (CO_2 certificates) can be traded on stock markets. They are then subject to the powers of supply and demand on the financial markets, and they become the subject of speculation. Their rational is the attainment of as high a return as possible, not that of as great a reduction in greenhouse gas emissions as possible in order to stabilise the climate.

In capitalism in general and as a result of the liberalisation of global financial markets in recent decades in particular enough innovative securitisation instruments have been developed to enable both a growing volume of paper oil to be dealt with and to handle the certificate trade with CO_2 pollution rights. On the one hand there is the airy, odourless and colourless CO_2, the concentration of which in the atmosphere measured in ppm is continually increasing and which is responsible for the greenhouse effect. On the other hand there is „paper CO_2" in the form of state-licensed rights to pollute the atmosphere with greenhouse gases, which can be sold as securitised certificates if they are not needed and which must be bought if they are needed. The emission certificates are an economic good „to which the principles of freedom of movement of goods apply" (according to the previous German black-red federal government, quoted from Financial Times Deutschland 21.4.2008). The circulating and partly securitised bonds, paper oil and emission certificates are compared with the returns which can be achieved with other securities on financial markets. Climate policy becomes the concern of profit-hunters. The market for emission certificates is booming, as authors from the World Bank write: from 2006 to 2007 alone the volume of worldwide traded emission certificates (99% of which took place within the European trading system) doubled from approx. US$ 25 bn to US$ 50 bn. Trade within the framework of the „Clean Development Mechanism", i.e. between the „Northern" industrial countries and the „Southern" developing countries (whereby China plays the major role in the „South"), has more than doubled, from US$ 6.5 bn to US$ 13.6 bn (Capoor/Ambrosi 2008). It is uncertain whether this growth can continue, however. This depends firstly on the design of the emission trading system (ETS) and secondly on the development of the financial markets during and following the financial crisis.

2   The Tragedy of the Atmosphere

Fossil primary energy is transformed into secondary and final energy and into the energy services such as heating, motion, lighting (cf. Wagner 2007: 29ff.). The losses of energy into the environment (e.g. the heating of cooling water, the explosion of oil platforms (such as Deepwater Horizon in April 2010), leakages from oil pipelines, tanker collisions, the „simply normal" pollution of the environment caused by the „non-conventional" exploitation of oil sand) and above all the deposits of combustion residues in the atmosphere and in other spheres of the Planet Earth should not be forgotten. In the case of the carbon cycle, forecasting uncertainties must also be taken into consideration. The chief economist at the IEA, Fatih Birol, admitted in an interview at the beginning of 2008 firstly, that the IEA's forecasts as a rule exaggerated the amount of oil reserves which could still be exploited and should therefore be corrected downward, and secondly that it was to be expected that the price of oil would remain permanently high. Thirdly, Birol doubted that „markets alone could solve the problem" (Birol 2008: 38). He concludes: „… I think we should leave oil before oil leaves us …" (Birol

2008: 41). However, the hope that the explicit abandonment of the previous line followed by IEA, to bring the final reserves of fossil fuels to the surface through massive investments in drilling equipment, transport and refineries, could herald the beginning of a sustainable energy regime was already destroyed by the IEA in June 2008. The IEA places its bets on the expansion of atomic energy instead of on renewable energy. It is of the opinion that not only „many thousands" of wind turbines are necessary, but also that 1300 new atomic reactors will be needed by the middle of the century. Every year 30 to 40 new reactors would have to be connected to the network somewhere in the world. In order to achieve this, about 45,000 billion dollars would have to be invested (cf. Süddeutsche Zeitung 7/8 June 2008). The structures of the fossil and nuclear energy system with large, centralised power station operators can be extended far into the future and with them the economic and political relationships of power based on them. If the black fossil-nuclear utopia becomes reality the conversion into a renewable energy system would be blocked for decades, unless this was designed on a large scale just like the fossil plants. Examples of this are megaprojects such as the photovoltaic and thermic production of electricity in the Sahara (Desertec).

Fossil and nuclear energies offer many advantages within the framework of the capitalist process of accumulation. From the individual point of view it would therefore be just as irrational to forgo the utilisation of these energies as it would be rational according to the global goal of stopping the increase in greenhouse gases in the atmosphere to opt out of the fossil energy regime. This contradiction between individual and societal rationality was addressed 40 years ago by Garrett Hardin as „the tragedy of the commons" (Hardin 1968): the pursuit of a behavioural logic based on individual rationality in the use of fossil fuels leads to an irrational overconsumption of the global commons, in this case the atmosphere of the Planet Earth. Therefore, there must be rules for the reduction of the emission of greenhouse gases into the atmosphere unless humanity has resigned itself to the motto „ après nous le deluge ".

Several conclusions can be drawn from the fact that there is a contradiction between individual rationality and societal irrationality. The first and most fundamental is: we start at the beginning of the energy chain and see to it that the fossil spirit stays in the bottle and that the CO_2 spirits which have already escaped are put back into it, if this is possible. Carbon should not end up in the atmosphere in the form of greenhouse gases, but where the carbon cycle began: in the caverns of the Earth's crust. This would only be possible if the fossil fuels are not extracted in the first place and if instead of the fossil reserves rays of solar energy are used, i.e. if the production and utilisation of renewable energy gain priority.

This proposal was made by the then Energy Minister of Ecuador, Alberto Acosta, in 2007 and taken up and officially presented by President Correa. The 920 million barrels of oil in the Ishpingo-Tambococha-Tiputini (ITT) field in the Yasuni Nationalpark should stay in the ground, thereby preventing CO_2 emissions to the tune of 410 million tonnes. (That would be equivalent to about 90% of the annual emissions of the German firms participating in emissions trading, which were given emission rights for 453 million tonnes in 2008). The neutralisation of the oil from the ITT field would also mean that the oil would not be valorised and commercialised by the oil companies and would not be able to circulate in the form of paper oil. The abstention from the extraction of the oil would therefore be nothing less than an act of capital annihilation, because the reserves have already been included into the capital stock of private and state companies. Ecuador is demanding compensation for this, „money from outside in recognition of its foregone monetary revenue" (Martinez-Alier/Temper 2007: 18). This may make sense as long as it remains an exception. It would not be suitable as a general rule for the compensation of non-extracted oil, however, for if there were to be monetary flows in compensation for non-extracted oil to corporations or states which left the oil in the ground, then the question is posed as to how and where the money could alternatively be invested as capital. Physical reserves of fossil fuel are also economic capital, which would be expropriated and annihilated to a corresponding extent by the renunciation of the valorisation of the reserves.

If the bottle is not corked, expectations will be directed towards the sinks in which CO_2 is bound and which are therefore listed in the Kyoto Protocol as CO_2 reduction (as emission reduction units, certified emission reductions etc. cf. Weistroffer 2007: 19f.) in the context of Joint Implementation (JI) and the Clean Development Mechanism (CDM) (cf. Capoor/Ambrosi 2008). The more CO_2 is bound in sinks in the one location (e.g. in the course of afforestation measures in the global South), the more CO_2 can be emitted in another location (e.g. in the global North). This may sound simple but the calculation of the effect of sinks on CO_2 emissions is difficult. The sinks are not designed for permanence, and they continually themselves become sources of CO_2, e.g. when afforested woodland is cleared. It is therefore an illusion to believe that CO_2 emissions can be compensated by the „project-based" mechanisms Joint Implementation and CDM. The carbon spirit remains active once it has escaped from the bottle. Nevertheless, by now a „run" on CO_2 sinks has been triggered because they generate emission rights, which can be sold with profit as an economic good, even if there is no scientific evidence of CO_2 absorption or only evidence of doubtful provenance (on the role of the IPCC in the study of „Land Use, Land Use Change and Forestry" cf. Lohmann 2006: 36). From 2006 to 2007 alone, the quantity of CO_2

emissions compensated for by the CDM and JI rose from 611 to 874 million tonnes of CO_2 equivalent, which represents an increase in the value from US$ 6.5 bn to US$ 13.6 bn (Capoor/Ambrosi 2008: 1). This again shows the doubling of the energy chain: the quantitative increase, at 43%, is high, but it remains far below the increase in the capital value of 109%. Once an emission right has been securitised and become tradable in the form of a certificate only the returns are important.

If fossil fuels cannot be left in the ground and if the assumption of effective sinks for the absorption of CO_2 is an illusion, a third question is posed: can the CO_2 be captured and stored safely in the earth („CO_2 capture and storage" – CCS)? It is in fact a fascinating idea to return the emissions from carbon combustion to the caverns of the Earth's crust from which the hydrocarbons (lignite, coal, oil and gas) were originally extracted. Other ecosystems would be safe from the CO_2 emissions, and there would be no greenhouse gas effect above and beyond that which has been provoked by the emissions to date. The emissions of CO_2 in the energy sector would continue to take place but they would not increase the concentration of greenhouse gases in the atmosphere and would thus also have no effect on the radiation budget of the Earth, would be „neutral" with regard to the greenhouse effect. We must pour water into this wine, however. Up to now CCS is simply an idea, the realisation of which is doubtful, and even if it is mature enough to be put into practice by 2020 at the earliest (estimation by the WBGU 2003 and the energy supply companies) then the greenhouse effect will possibly already be too far advanced and the solution may have arrived too late. Besides, even if the CCS did function, this would only be an effective incentive for the construction of further coal power stations. The old structures of energy supply would be strengthened. The incentive for abandoning fossil energy supply would be lost.

Furthermore, it must be taken into account that CCS could at best offer a solution for the energy sector, not for the emissions caused by road traffic. Other solutions must be found for that. There is unlikely to be an emission-free car in the coming decades. Therefore, traffic systems must be redesigned in order to guarantee the necessary mobility.

Since nuclear energy also offers no solution to the energy and climate problem (cf. WBGU 2003; Forum für Atomfragen des österreichischen Lebensministeriums 2007), climate change remains, as Tony Blair warns, „probably the single most important issue we face as a global community" (quoted in Lohmann 2006: 23).

3   Should the Pollution in the Atmosphere be Reduced by Means of Regulative Power or by „Market-based" Instruments?

The „issue" of ominous climate change can in principle be dealt with (1) by political and military power, (2) in line with the market economy using „market-based instruments", or (3) according to the principle of global solidarity. Only strong actors, e.g. strong nation states such as the USA or alliances such as NATO, which have accumulated a considerable amount of economic and political power in the past (precisely with the aid of fossil fuels), can enforce their interests over others in the dispute over access rights to the hydrocarbon reserves and over rights to the pollution of the atmosphere. That would be pure *„grandfathering"*: those who have availed themselves of many environmental resources in the past and have thus gained power, can continue to do so in the future. This is risky, however, because of the inevitably confrontational policy, which provokes opposition (e.g. from social movements and from actors operating „asymmetrically"). New powers, e.g. China and India, would be forced to take part in the geopolitical dispute over energy sources and emission rights.

It is therefore the obvious alternative to follow the *second* principle and make use of the market mechanism. The idea arises of trading pollution rights among market actors, in order to get away from the taxing by the state of individual welfare gains which have arisen due to environmental pollution and hence lead to the diminishing of the welfare of society as a whole („Pigou tax" – Pigou 1960), and to move towards acceptable negotiations among market participants on the damage and its monetary compensation (Coase 1960; cf. the contribution by Nell/Semmler/Rezai in this volume). Private, market-based self-regulation should replace political control by the state.

This is not as self-evident as it sounds, however. Reality must also force us to think, and it has been doing this emphatically since the liberalisation of global financial markets in the 1970s. Following every financial crisis liquid capital searches more intensively for alternative investment opportunities, which should yield returns which are as high as possible. In this connection one is reminded in the negotiations on the Kyoto Protocol of the idea of Ronald Coase (1960) and others, not to avert the climate catastrophe with taxes and other state interventions but to leave this to private actors on a market to-be-established. Just as the oil reserves are valorisable resources, the atmosphere as a depository for CO_2 emissions can be valorised. Without the deposition of CO_2, namely, the system based on fossil fuels would collapse and therefore the atmosphere is a „factor of production" in the transformation of fossil fuels into useable energy which simply for physical reasons cannot be done without, and which therefore must be purchased on the market in return for payments (Lohmann 2006: 55ff.), if and when rights of ownership over it have been established. Thus, it comes to pass that the actors in financial markets participate in decisions on energy and climate policy.

The establishment of a system of emissions trade takes place in several steps: *first,* the atmosphere has to be transformed from a global commons into a private good by a sovereign act of the states. The „pluriversum" of nation states has to find an international agreement on the state of the atmosphere. The Durban Declaration of African NGOs on the Kyoto mechanisms states, „In an atmosphere of privatisation they privatised the atmosphere" (quoted from: Hänggi 2007). Of course, no private property rights are granted (that would be impractical) but, rather, state rights to pollution of the atmosphere and an international agreement of the participating states in order to distribute the emissions rights. In this way no full property rights arise, but documented (certificated and securitised) utilisation rights which can be traded (allowances, permits, pollution rights; on the differences cf. Lohmann 2006: 71-86). In this way economic goods are created which become part of the capital value of a firm. Profitable transactions can now take place on specialised financial markets.

Secondly, it must be clarified who is to be given the pollution rights. After all, CO_2 is emitted by all living beings, humans, animals and plants. Humans breathe in oxygen, among other things, and breathe out carbon dioxide, among other things, about 700 grams of CO_2 daily (Wagner 2007: 181). That is a natural process. In addition to their personal emissions caused by breathing all humans are given rights to „personal pollution contingents ... as an economic instrument for the limitation of greenhouse gas emissions" (Rahmstorf/Schellnhuber 2007: 128). It would only be rational if these were also extended to solid waste, the use of bottles, etc. in the sense of an „ecological footprint". A value of 3 tonnes per capita is regarded as „fair", compared to the 12.6 tonnes which the citizens of the industrialised countries emit per capita per annum and *vis-à-vis* the developing countries with an average of 2.3 tonnes – ignoring the differences within the individual country groups and among the different classes (Santarius 2007). Unneeded rights are offered on the market for „domestic tradable quotas" (DTQs) and those who wish to utilise more emission rights have to buy them.

CO_2 sinks can turn into a gold mine because they enable selling CO_2 rights, e.g. to those who emit more CO_2 than the quantity allowed. It is hoped that the trade with DTQs will lead to an equalisation of welfare between the North (where people buy additional emission rights) and the South (where emission rights can be sold) and that in this way one millenium development goal of the UNO, the reduction of poverty, can be achieved (Santarius 2007). And all of this without touching the functional conditions of the global capitalist system, without regulating the financial markets, simply by politically creating certified rights which can be traded globally as securities on stock exchanges or „over the counter" (OTC) and thus trigger financial flows. Rahmstorf/Schellnhuber, who are not disinclined towards the idea of personal pollution rights, summarise: „The approach is of course not yet ready to be put into practice but it opens up new and interesting perspectives" (Rahmstorf/Schellnhuber 2007: 128). Rahmstorf/Schellnhuber are wrong. The approach is theoretically inadequate and will therefore never be ready to be put into practice.

Upper limits (caps) must now, *thirdly*, be set for the firms participating in emissions trade. This creates an artificial scarcity. Nobody would have to buy emission certificates if there were plenty of them available because caps did not exist or were not restrictive enough. The „caps" are necessary if emissions trade is to function at all. For this reason, *fourthly,* a trading platform must be created on which the pollution rights can be traded. Since the emissions are a global problem this platform must in principle be open to owners of certificates from all regions of the world. This opens up interesting fields of business for CO_2 brokers and traders (Sinai 2006), including large banking houses which earn good money from the fees. Financial investors are interested in the returns which they can attain from the *trade* in the certificates, not in the *cap*, and certainly not in lowering it, because then the supply of certificates would fall and in spite of increasing prices an expansion could not take place because the supply – ceteris paribus – could not be expanded. The less that can be traded because the caps for CO_2 reduction are set very restrictively for climate policy reasons, the more difficult it will be to do business – and vice versa. Emissions trade therefore brings actors into play who strive after profits from financial speculation and who have no interest in the control of greenhouse gases. This again shows that the rationality of the market, which unconditionally demands upper limits to pollution because otherwise there would be no scarcity, and the rationality of the individual market actors, who have an interest in an allocation of pollution rights which is as generous as possible, are not the same.

The double character of real atmospheric pollution and the trade with pollution rights (*paper emissions*) offer many opportunities for crooked deals, including the double game played by the USA: they have not signed the Kyoto Protocol, which would oblige them to commit to reducing real atmospheric pollution (even though to an inadequate degree). Firms from the USA wish to participate in the trade with pollution rights, however, because they can earn money by doing so (cf. „Nasdaq steigt in Klimahandel ein" (Nasdaq enters climate trade), in: FTD 13.5.2008). Financial investors can force their way into the new politically created markets in order to offer innovative and profitable financial instruments. In this way, part of the over-accumulated global liquidity can be invested.

Emissions trade, which was designed for the reduction of CO_2 emissions, is transformed into a new vehicle of financial speculation – and even of fraud. In the „Human Development Report" of 2007 it is pointed out that „cap-and-trade schemes are open to manipulation by vested interests. As one commentator has written, issuing allowances is ‚in essence printing money for those in control of the permits'. Who gets how many permits and at what price are issues that have to be determined through political processes. Inevitably those processes are open to powerful actors' influence — power companies, oil companies, industry and retailing, to name a few. Pandemic cheating has been highlighted as the Achilles' heel of cap-and-trade approaches" (UNDP 2007: 14; cf. also Lohmann 2006: 58-63). If the stock exchanges and the financial markets go into crisis then oil and CO_2 papers are not exempted. Not only did it already come to the fraudulent practices in the allocation of pollution rights mentioned above, above all within the framework of the „Clean Development Mechanism", but it is also possible to obtain certificates for CO_2 emissions at low cost which can, for example, be sold to power station operators in the industrialised countries (cf. e.g. FTD 11./12./13.12.2009: Milliardenbetrug im Klimahandel (Billions fraud in climate trade); FTD 14.12.2009: Rauchzeichen (Smoke signals), p. 23; Der Spiegel 7.12. 2009: Die Klima-Mafia (The climate mafia), p. 90; SZ 29.4.2010; p. 11: Razzia gegen Betrug mit Klimazertifikaten (Raid against fraud with climate certificates), p.5; FAZ 29.4.2010). In the trade on the exchanges with papers which have been „originated" in a dubious fashion the frauds have obviously been perpetuated, as Europol criticises (cf. FTD 29.4.2010; FAZ 29.4.2010).

The *fifth* step is the allocation of pollution rights to the firms engaged in their trade. Will they be allocated free of charge and if so in what quantities, or will they be sold for money, e.g. by being auctioned? If the pollution rights are allocated generously then they will not be worth the paper on which the right to emit CO_2 is documented. If the certificates are not scarce they will not become a tradable good. The National Allocation Plan for Germany (until 2007) intended e.g. that CO_2 emissions should be reduced by 2 per cent. That is much less than laid down by the Kyoto objectives or the 450 ppm scenario demanded by the IEA. The power station operators and other branches subject to emissions trade require fewer rights than have been allocated to them. The caps have been set so that the supply is large and demand is low. Therefore the price for the emission of one tonne of CO_2 collapsed from its historical high of approx. 30 euro to a few euro (€8.60) and finally to one euro per tonne of carbon dioxide (cf. Nell/ Semmler/Rezai in this volume). Of course, the supporters of emissions trade regard this simply as a technical problem which could be solved by stricter emission limits in the allocation of pollution rights.

The firms which possess emission rights can even make a profit without selling the certificates on the market. If they integrate the price of the pollution rights into the price of electricity, which has happened within the framework of the ETS, profits rise without the firms participating in the trade of certificates („windfall profits"; on this cf. Brouns/Witt in this reader). They calculate an assumed certificate price (of 30 euro/t) into the electricity price which they demand from their customers, although they have paid much less for a tonne of CO_2. Even if it is accepted that these opportunities are a defect and a disadvantage of the EU's Emissions Trading System (ETS) and that they should be „corrected", the fundamental principle of emissions trade is not affected and neither is the doubling of real CO_2 emissions and the tradable rights to them.

Sixthly the process must be subjected to monitoring in order to guarantee transparency and avoid fraud. But who is to be responsible and which role will environmental movements and their research institutes play?

Seventhly, it must also be decided who is to receive the returns from the auctioning of certificates and for what purpose these are to be used. After all, in Germany as of 2013 between 6 and 10 bn euro could be attained annually by means of auctions. If this money is directed into growth-producing investments or into consumption then this will trigger a further use of fossil fuels, so that the problem which was supposed to be solved will actually get worse. Only the „neutralisation" of the returns in order to close the bottle from which the fossil spirit is flowing, i.e. the transition to renewable solar energy, would be a solution. But there is no incentive to put a cork in the mouth of the bottle from which the spirit of fossilism is escaping into the carbon cycle, to develop alternative forms of energy and to save energy. Below the cap the traditional energy policy line can be continued. There is no reward for changing to another energy policy line of development.

Although the defects of emissions trade are evident, even environmental organisations are strongly in favour of it. Larry Lohmann shows how internationally operating NGOs, together with environmental associations and the World Bank, have warmed to the idea of emissions trade, from the World Resources Institute and Greenpeace to the UNICE of European entrepreneurs and the World Wide Fund for Nature (Lohmann 2006: 58-60). In Germany there are supporters of emissions trade among the members of the Green Party and in many NGOs, e.g. Germanwatch. The „climate alliance" of German NGOs came out in favour of emissions trade in a position paper in April 2007, although only if the emission rights are auctioned (http://www.die-klima-allianz.de/position.php) and in the context of other political measures for the reduction of CO_2 emissions. Many NGO representatives regard the market as an adequate problem-solver for the contradictions of industrialised society and for the energy and climate crisis. It is assumed that the individual rationality of the certificate traders, i.e. their private striving for profit, and the collective rationality of society, i.e. a considerable reduction in the emission of greenhouse gases, are not only compatible, but that in addition the collective rationality prevails in the behaviour of individual actors. Admittedly, some cracks have appeared in this belief during the financial crisis, which also caused setbacks in the trade in certificates.

4   A few Conclusions

The number of those who doubt the sense of emissions trade is growing. In the *Financial Times* (25th of April 2007) even before the financial crisis emissions trade was described as a modern form of the sale of indulgencies in the Late Middle Ages. Its efficiency is doubted. As a result of the trade in certificates it is more profitable for energy companies to replace old coal power stations, for example, with new ones (which continue to emit CO_2 for several decades, even if they do so in a „more efficient" way than the old ones) than to turn onto the path of renewable energy (von Fabeck 2007).

Emissions trade is designed not to challenge the institutional foundations of the capitalist system, although the inherent dynamism of capitalism contributes to the overuse of natural resources and to the destruction of ecosystems. Looking at the carbon cycle in isolation or wanting to optimise the energy chain with the aid of market-based instruments is not enough to overcome the menacing climate problem and the food crisis it brings in its wake. The doubling of hydrocarbons in fossil fuels and in a capital value must be taken into account if they have been valorised by being extracted and brought to the surface of the Earth and are now not only part of the carbon cycle but also of a valorisation cycle.

In the discussion on climate protection, however, the accumulation dynamism of capital, the natural limits, are not taken into account, instead they are ignored. Reduction scenarios such as that of the Stern Report (Stern Review 2006) or that of the IPCC (IPCC 2007) always assume that the goals can be achieved with the means of emissions trade and that a „*win-win*-situation" can be established without affecting the foundations of climate capitalism (Newell in this volume): climate protection is possible even if (or precisely if) economic growth continues. The OECD (2008) has calculated that a little more than 1% of global GDP would be enough for climate protection in 2030 in order to avoid the losses otherwise to be feared of up to 20%. This is stated by the Stern Review (2006). The costs of 1% result in a reduction of the rate of growth of 0.03% on a yearly average, so that GDP in the year 2030 will only be greater than today by 97% instead of approx. 99%. Precisely this neoliberal optimism must be doubted as it irrationally assumes that it is possible to reduce CO_2 emissions at the end of the energy chain without limiting the combustion of fossil fuels at its beginning.

The doubts with regard to market rationality were already expressed by Karl Polanyi in his criticism of „disembedded" markets which he wrote in the 1940s (German translation: Polanyi 1978). Markets require regulation which contributes more than simply the seven steps towards the emissions trade platform described above. On financial markets emission certificates, with which the CO_2 emissions are to be reduced, are traded as assets which are intended to provide a return. Once the certificates have been sold for the first time the relationship to the original obligation of reduction is invisible. Trade with them is directed exclusively towards attaining greater returns than from other, alternative investments.

A CO_2 tax or a tax on fossil fuels could in fact have more effect than emissions trade (cf. also Nell/Semmler/Rezai in this volume), not least because the profits from emissions trade partly flow, in the form of „windfall profits", to private enterprises which can only use them to speculate, while a tax could be used according to political criteria with democratic participation, although only in conjunction with a large bundle of further measures such as stricter regulations for the utilisation of energy, punishment for polluters, the end of subsidies for fossil fuels and the abandonment of the billion-dollar World Bank projects for the exploitation of fossil energy (cf. for example Smith 2007: 127).

In the final analysis it will only be possible to stabilise the climate if the energy system is reconstructed, if the fossil fuels remain in the earth and if renewable energy is used in their stead. This requires a sustainable change in production structures and patterns of consumption throughout the world, in a variety of quite different ways suited to the different natural conditions, economic stages of development and cultural traditions.

This brings us to the third possibility to reduce greenhouse gas emissions mentioned above, in addition to the use of political and military power and the application of market-based instruments. Limits to emissions (caps) can be laid down collectively and in solidarity. They are not tradable but represent a commitment to radically limiting the supply from the fossil reserves in the Earth's crust, i.e. to corking the bottle, by transforming the energy system as a whole in the direction of renewable energy sources. This will probably only be possible if the world of work is also changed toward an economy based on solidarity.

References

Altvater, E. (1992): Der Preis des Wohlstands oder Umweltplünderung und neue Welt (un)ordnung. Münster: Westfälisches Dampfboot.

Altvater, E. (2005): Das Ende des Kapitalismus wie wir ihn kennen. Münster: Westfälisches Dampfboot.

Anders, G. (1956/1992): Die Antiquiertheit des Menschen, 2 volumes. Munich: H.C. Beck.

Brunnengräber, A. (2011): Multi-Level Climate Governance. Strategic Selectivities in International Politics, in: Knieling, Joerg, Leal Filho, Walter (2011): Climate Change Governance, HafenCity University Hamburg, Frankfurt: Springer (forthcoming).

Birol, F. (2008): Interview by Astrid Schneider with the title „Die Sirenen schrillen". In: Internationale Politik, April 2008: pp. 34-45.

BMU (2006): Bundesministerium für Umwelt, Naturschutz und Reaktorsicherheit: Die projektbasierten Mechanismen CDM & JI – Einführung und praktische Beispiele, Reihe Umweltpolitik, 2nd ed., November 2006. Available at www.bmu.de/files/ pdfs/allgemein/application/pdf/broschuere_cdm_ji.pdf, accessed February, 4th 2011.

BMU (2007): Bundesministerium für Umwelt, Naturschutz und Reaktorsicherheit, Hintergrundpapier: Emissionshandel in der zweiten Handelsperiode 2008 – 2012 Einfacher – wirksamer – effizienter. April, 18th 2007.

Capoor, K.; Ambrosi, P. (2008): State and Trends of the Carbon Market 2008, The World Bank. Available at http://carbonfinance.org/docs/State___Trends--formatted_06_ May_10pm.pdf, accessed February, 4th 2011.

Enquete Kommission „Schutz der Erdatmosphäre" des Deutschen Bundestages (ed.) (1994): Schutz der Grünen Erde – Klimaschutz durch umweltgerechte Landwirtschaft und Erhalt der Wälder. Bonn: Economica Verlag.

Fabeck, W. von (2007): Emissionshandel kritisch bewerten – Neue Aufgaben der Umweltverbände. In: Solarzeitalter, Vol. 19, No. 2 (2007): pp. 42-52.

Gabriel, S. (2008): Vorwort. In: Schüle, Ralf (ed.): Grenzenlos handeln? Emissionsmärkte in der Klima- und Energiepolitik. Munich: Oekom-Verlag, pp. 7-8.

Georgescu-Roegen, N. (1971): The Entropy Law and the Economic Process. Cambridge (Mass.), London: Harvard University Press.

Hänggi, M. (2007): Gerecht geht nicht, in WOZ die Wochenzeitung. March, 29th 2007.

Hardin, G. (1968): The Tragedy of the Commons. In: Science, No. 162 (1968): pp. 1243-1248.

Heymann, E. (2007): EU-Emissionshandel. Verteilungskämpfe werden härter. In: Deutsche Bank Research, January, 25th 2007.

IEA (2007): International Energy Agency: World Energy Outlook 2007. China and India Insights. Paris.

IEA (2009): International Energy Agency: World Energy Outlook 2007. China and India Insights. Paris.

IPCC (Intergovernmental Panel on Climate Change). 2007. Fourth Assessment Report of the IPCC (2007) on Climate Change. Available at http://www.ipcc.ch/publications_ and_data/publications_and_data_reports.shtml, accessed April, 11th 2011

Kromp-Kolb, H.; Formayer, H. (2005): Schwarzbuch Klimawandel. Wie viel Zeit bleibt uns noch? Salzburg: Ecowin.

Lohmann, L. (guest editor) (2006): Carbon Trading. A critical conversation on climate change, privatisation and power. In: Development Dialogue No. 48, September 2006.

Lohmann, L. (2006a): Carry on pollution. Comment and analysis in New Scientist, December, 2nd 2006. Available at http://www.thecornerhouse.org.uk, accessed February, 4th 2011.

Martinez-Alier, J.; Temper, L. (2007): Oil and Climate Change: Voices from the South. In: Economic & Political Weekly, December, 15th 2007: pp. 16-19.

OECD (2008): OECD Environmental Outlook to 2030. Paris: OECD.

Polanyi, K. (1978): The Great Transformation. Frankfurt am Main: Suhrkamp.

Rahmstorf, S.; Schellnhuber, H.-J. (2007): Der Klimawandel. Munich: C.H. Beck.

Santarius, T. (2007): Klimawandel und globale Gerechtigkeit. In: Aus Politik und Zeitgeschehen, No. 24, 2007.

Schafhausen, F. (2007): Der Emissionshandel, das unbekannte Wesen. In: Müller, Michael; Fuentes, Ursula; Kohl, Harald (eds.): Der UN-Weltklimareport. Bericht über eine aufhaltsame Katastrophe. Cologne: Kiepenheuer und Witsch: pp. 377-386.

Sinai, A. (2006): Tausche Wasserkraft gegen Treibhausgas. Was vom Kioto-Protokoll übrig blieb. In: Le Monde diplomatique, deutsche Ausgabe, January 2006: p. 20.

Smith, K. (2007): Klimawandel und Emissionshandel. In: Melber, Henning; Wilß, Cornelia (ed.): G8 Macht Politik. Wie die Welt beherrscht wird. Frankfurt am Main: Brandes & Apsel.

Stern, N. (2006): Stern-Review on the Economics of Climate Change, Her Majesty's Treasury. Government of the United Kingdom. Available at http://www.hm-treasury. gov.uk/independent_reviews/stern_review_economics_climate_change/sternreview_in dex.cfm, accessed February, 4th 2011.

UNDP (2007): Human Development Report 2007/2008. Fighting climate change. Human solidarity in a divided world. New York.

Wagner, H.-J. (2007): Was sind die Energien des 21. Jahrhunderts? Der Wettlauf um die Lagerstätten. Frankfurt am Main: Fischer Taschenbuch Verlag.

WBGU (2003): Wissenschaftlicher Beirat der Bundesregierung Globale Umweltveränderungen: Welt im Wandel. Energiewende zur Nachhaltigkeit. Berlin, Heidelberg, New York etc.: Springer.

Weistroffer, C. (2007): Klimawandel bewältigen. Die Rolle der Finanzmärkte. In: Deutsche Bank Research, September, 24th 2007.

World Bank (2007): State and Trends of the Carbon Market 2007. Washington D.C., May 2007.

A Brief History of Emission Trading Systems

Miranda A. Schreurs

Introduction

There is a growing body of experience in the United States and internationally with emission trading systems (ETS). On the whole, existing systems suggest that there is an important role to be played in pollution control by ETS, but a system's effectiveness depends on the nature of the pollutant, the system's design, the range and kind of facilities covered by the system, and the ability of a central authority to effectively monitor facilities' emissions and compliance with the rules of the trading system. Depending on the nature of a pollutant, emission trading systems can prove preferable to strictly command and control approaches to pollution control as they provide firms with greater flexibility in determining when and how to reduce a pollutant. There are, however, some environmental justice issues with ETS that must also be considered.

Emission trading systems are often referred to as cap and trade systems. They have as their main purpose the control of a pollutant through the allocation of permits (also known as allowances) to polluting firms. Each permit represents the amount of the pollutant that a facility may emit. Firms are not allowed to operate unless they have sufficient permits to cover the pollutants they discharge. Firms that do not comply are subject to heavy fines.

The underlying logic behind emission trading systems is that because a central authority limits the number of allowances in the system, allowances take on value. Essentially, the permits put a price on the right to pollute. Market signals, thus, can help a firm determine whether it is more cost effective to continue polluting or to reduce their pollution load thereby reducing the number of allowances they must obtain to operate. As long as the price of the allowances is high enough, this approach pushes firms that can most cost effectively reduce pollution to do so.

In the language of cap and trade systems, a cap refers to a maximum emission (or production) level beyond which total emissions (or production) are to be prevented. The units (e.g. firms, facilities, power plants) governed by an emission trading system are either allocated (e.g., based on historical emission levels) or emission allowances must be purchased (e.g. through an auction). Caps and the units to be governed by the system are usually determined through some form of political bargaining but with a central authority ultimately determining which firms are to be covered by a system and at what level a cap is to be established.

In an emissions trading system, units can buy emission and sell emission allowances. A firm desiring to expand its operations (and as a result, its emissions) will need to either improve the efficiency of its existing operations, bringing down emissions in the process so that it can expand its operations and stay within the level of emission allowances it holds or obtain emission allowances by buying them from another firm. Firms that can save money by cutting their emissions and selling the allowances (also known as emission credits) they hold will choose to do so.

In order to reduce pollution over time, ETS usually build in some kind of mechanism that will lead to a reduction in the number of allowances in the system. This can either be done by a central government lowering the level of the cap over time or through a „retirement" system whereby a certain percentage of allowances are taken out of the system with each trade that occurs. Theoretically, over time this pushes up the price of remaining permits. Non-system actors, such as non-governmental organizations may also choose to purchase permits in order to retire them from the system. Alternatively, a corporation may choose to retire a share of its emission allowances, donating them, for example, to a non-profit organization. The end effect is to drive up the cost of polluting giving firms a greater incentive to reduce their pollutant emissions.

An alternative to the cap and trade model is a baseline and credit approach. In this kind of system, no total cap is established. Instead, credits are created when a firm brings its pollutant below a baseline (e.g. emission levels at time X). These credits can than be sold to other firms that are exceeding their own baseline.

Already in the 1960s, when the United States began developing national environmental pollution control regulations, economists were making a case for the efficacy and flexibility of emission trading systems. It was not until the 1970s and 1980s, however, that emission trading systems began to be employed by the United States. Europe began experimenting with emission trading systems in the 1980s and 1990s. Below various emission trading systems and the innovations they have led to are briefly introduced.

Emissions Trading Systems

The Environmental Protection Agency's Emission Trading Program under the Clean Air Act Program to improve local air quality (1974)

Developed in the 1970s, the U.S. EPA's Emissions Trading Programs were the first in the world to incorporate emissions trading systems in an effort to address air pollution. There were several programs developed. A „netting" system allowed new large sources to be exempted from review procedures if existing emissions elsewhere in the same facility were adequately reduced. In 1976, an „offset" program was added to the system. Companies wanting to establish a new facility in an area that was not in compliance with the national ambient air quality standard could do so if they could reduce existing emissions at another facility within the non-attainment area by at least the same amount as the new facility would emit. A bubble system was also formulated. This allowed a firm to aggregate all of its emissions from its various facilities when calculating compliance with emission levels. Finally, a banking system was started that allowed firms to obtain credits for future use when they took actions that reduced their emissions below a specified standard. While pioneering in its approach, the program was not widely used as it was voluntary and highly complex in its design. It required extensive reporting and compliance procedures limiting firm ability/willingness to employ the system. According to one assessment, the programs „constitute the first official recognition of the potential value of emissions trading, but the disappointing experience with these programs is the primary reason for the early reputation of emissions trading as a theoretically desirable but largely impractical concept."[1]

Lead Trading Program

In 1973, the U.S. EPA initiated a „phasedown" program to bring the levels of lead in gasoline down to 0.5 grams per gallon by 1980 in large refineries and by 1982 in small refineries. Refineries could average their total (both leaded and unleaded) output to reach the 0.5 standard. In 1982, the standard was changed to 1.10 grams per leaded gallon but eliminated the provision that allowed averaging.[2] At the same time, the EPA introduced a lead trading program for gasoline

---

[1] A. Denny Ellerman, Paul L. Joskow, and David Harrison, Jr., „Emissions Trading in the U.S.: Experience, Lessons, and Considerations for Greenhouse Gases," Pew Center on Global Climate Change. www.pewclimate.org/global-warming-in-depth/all_reports/emissions_trading/emissions_execsumm.cfm.

[2] EPA Press Release, „EPA Sets New Limits on Lead in Gasoline," March 4, 1985.

refineries. A refinery could exceed the lead content limit in its own gasoline if it purchased an equivalent number of allowances from another refinery that had reduced its lead content below the level required by the EPA.

A major innovation with this system was its allowance of banking. In 1985 the EPA further tightened its lead content rules. For mid-1985 the level was set at 0.5 grams per leaded gallon and for 1996 at 0.1 grams per leaded gallon. To accompany this change in the regulation, starting in 1985, refineries were allowed to bank credits that they accumulated for meeting the phase down requirement early and use them in subsequent years. The program was extensively used from 1985-87, when it was terminated because lead phase down was complete. The EPA estimated that savings attributable to the program were significant and the phasedown of lead was sped up by the trading program.

Trading in the Right to Produce and Consume Ozone Depleting Substances under the Montreal Protocol

One of the first international ETS was created by the Montreal Protocol on the Control of Substances that Deplete the Ozone Layer. The Montreal Protocol created schedules for the phase-down (and eventual phase-out) of several ozone depleting substances (ODS), including chlorofluorocarbons (CFCs). The phasedown and phase-out schedules were different for industrialized and industrializing countries. When the system was introduced there were only 17 ODS producers in the world. By 1997 there were 79 producers (57 of which were in developing countries).[3]

Under the Montreal Protocol a provision was added to allow „industrial rationalization" of ODS production and consumption rights. Essentially, this created an international emissions trading system. The goal was to encourage a more cost efficient reduction of the production and use of chloroflourocarbons, consistent with the phase out plan established by the protocol. Firms that could most cost effectively reduce their production or consumption of ODS had an incentive do so and then sell their ODS production or consumption rights to another firm (possibly in another country) that needed them.

In the United States the government established ODS limits for all ODS producers based on their 1996 ODS production levels. The limits declined over time in accordance with the Montreal Protocol phase out schedule. Firms that wished to produce or use CFCs had to have an „allowance" issued by the U.S.

---

[3] Organization for Economic Cooperation and Development, „Lessons from Existing Trading Systems for International Greenhouse Gas Emission Trading, Annex I Expert Group on the United Nations Framework Convention on Climate Change Information Paper, ENV/EPOC(98)13/REV1, August 3, 1998.

Environmental Protection Agency. For every domestic trade that occurred, the EPA retired 1%t of the amount of ODS represented by the allowance from use. This was done to ensure that trading would lead to greater reductions than would occur without trading.

The U.S. EPA recorded 172 trades in 1992 of which only 1 was international. In 1994 there were 147 trades of which 9 were international. By 1995 the number of trades started to decline as production was declining based on the Protocol's phase-out time line. Monitoring was conducted by national governments based on records of ODS production, imports, and exports. In the United States all trades had to be reported to the EPA before they occurred. International trades required the notification and approval of all relevant authorities of the different countries concerned. The U.S. EPA, for instance, would only allow an international purchase by a U.S. firm to occur if the embassy of the selling firm's country declared that the country had reduced its production rights by the amount transferred.[4] The ODS trading system reduced the administrative costs of implementing the Montreal Protocol.

In the European Union, the European Commission allocated production quotas to ODS producers based on their historic production levels. European producers were allowed to trade within the European Union or with any other party to the Montreal Protocol as long as their combined production quotas did not exceed their combined quota limits. A similar system was put in place establishing ODS consumption quotas. Trade with non-parties to the Montreal Protocol were prohibited.

Developing countries often lacked the necessary monitoring capacity to comply with the emission trading rules of the Montreal Protocol. To aid them in their monitoring efforts, the Montreal Protocol's Multilateral Fund provided funds to help them establish National Ozone Units. Trade between non-European Union countries had to be reported and approved by the United Nations Enviromental Programme's Ozone Secretariat.

While there was some abuse of this system and some illegal trading, on the whole the system appears to have functioned reasonably well. Countries found in non-compliance by the Implementation Committee that was established under the Montreal Protocol were first provided with technical and financial assistance to help bring them into compliance but when this did not function could either be issued cautions or ultimately be suspended from the Montreal Protocol. Suspension would mean the loss of access to the ODS market of other Parties and loss of the right to financial support.

---

[4] Ibid.

NOx Trading Scheme under NOx Ozone Transport Commission

The 1990 Clean Air Act Amendments established the Ozone Transport Commission (OTC) to help states in the Northeast and Mid-Atlantic region to meet the National Ambient Air Quality Standard for ground-level ozone. Initially, the OTC's focus was on achieving year-round, region-wide emission limits based on Reasonable Available Control Technology.

Then in 1994, the OTC states took the initiative and signed a Memorandum of Understanding formulating a multi-state cap and trade program (the NOx Budget Program) to control NOx emissions and address regional transport of ozone. The program set a regional „budget" (a cap) on NOx emissions from power plants and other large combustion sources during the time of year when these states tend to be out of compliance (May 1 through September 30). The program covered 1,000 large combustion facilities. Under the program's cap and trade provisions, each state allocated emission allowances to their combustion sources in accordance with the portion of the regional budget allocated to the state.[5] Each allowance permitted a source to emit one ton of NOx during the ozone season. Firms could sell unused allowances or bank them for future use. Regardless of the number of allowances held by a source, it was not allowed to emit at levels that would violate emission limit requirements. The program was a collaborative state/federal partnership. The states established the program requirements and emission budgets and then the EPA administered data systems used to manage the program and provided technical assistance to the states in tracking allowance transfers, maintaining unit and account information, assisting with monitoring, and preparing annual reports. The program was highly successful, leading to a significant reduction in NOx emissions in the Northeast and Mid-Atlantic regions. According to the OTC, the trading program and the earlier Reasonably Available Control Technology requirements brought ozone season emission to approximately 60% below 1990 level emissions by the early 2000s. The Budget Program, moreover, was found to be cost effective.

The NOx Budget Program was replaced in 2003 by the NOx Budget Trading Program in response to the EPA's call for State Implementation Plans (SIP) to reduce the transport of ozone over broader geographic regions. The program expanded the number of states involved in trading, thereby expanding the electric power and industrial combustion sources covered by the program. Under the NOx SIP call program reductions were mandated in two phases. The first phase that went into effect in 2003 required further reductions in NOx emissions (of

---

[5] Ozone Transport Commisssion, „NOx Budget Program, 1999-2002 Progress Report," http://www.epa.gov/airmarkets/progress/docs/otcreport.pdf.

approximately 35-40%) from the states that had been involved in the earlier NOx Budget Program and in 2004 of designated facilities in the new states covered by the program.

1990 Clean Air Act Amendments: SO_2 allowances

Perhaps the best known example of an emissions trading system is the SO_2 emissions trading system established by Title IV of the 1990 Clean Air Act Amendments. The 1990 Clean Air Act required total U.S. emissions of SO_2 to be capped at around 9 million tons per year. This was to be achieved through a two phase process, the first starting in 1995 and the second in 2000.

In the first phase, the EPA assigned emissions limits to 263 of the most SO_2 emissions intensive generating units at 110 power plants operated by 61 electric utilities. The EPA allocated each utility a certain number of allowances based on its heat input during the baseline period (1985-87). Phase two extended the program to other fossil-fuel electricity generating facilities with total SO_2 emissions to be capped at approximately 9 million tons (an average of about 1.2 pounds of SO_2 per million Btu). The overall impact of the program was expected to bring emissions down to about half of what they were in 1980.[6]

Sources were issued tradable allowances. Each represented the right to emit one ton of SO_2. Facilities had to surrender an allowance for every ton of SO_2 emitted. Allowances that were not used could be traded or banked for future use. A small percentage of allowances (2.8%) were withheld from units and withheld for distribution through an annual auction to encourage trading and to make sure allowances were available for new electricity-generating units. The revenues from the auction were returned on a pro rata basis to the firms from which the allowances were withheld.

There was a particularly sharp reduction in SO_2 emissions in 1995, one year after the program went into effect. The reason for this was that firms had a strong incentive to reduce emissions beyond what was required and bank their use for future periods when marginal abatement costs were expected to rise (especially after 2000 with the start of Phase II of the program).

This program is considered one of the most successful examples of emissions trading and suggests that emissions trading can lead to sharp reductions in emissions at lower costs than would be the case with less flexible command and control regulations.

---

[6] A. Denny Ellerman, Paul L. Joskow, and David Harrison, Jr., „Emissions Trading in the U.S.: Experience, Lessons, and Considerations for Greenhouse Gases," Pew Center on Global Climate Change.   www.pewclimate.org/global-warming-in-depth/all_reports/emissions_trading/emissions_execsumm.cfm.

Carbon Emissions Systems: Corporate Initiatives

Corporations were the first to experiment with carbon emissions trading models. Multi-national corporations eager to find cost-effective means of reducing their own greenhouse gas emissions began experimenting internal emission trading schemes.

In 2000, TransAlta, a Canadian firm, for example, launched plan to eliminate greenhouse gas emissions by 2024, partly through emissions trading. In 2004, they announced the purchase of 1.75 million tons of greenhouse gas candidate Certificate Emission Reductions from Chilean agricultural company, Agrosuper. Their offset projects in the first years of operation included: gas recovery, energy efficiency, ruminant methane, landfill and coal mine gas to electricity, forestry, and soil sequestration.

BP launched an internal cap and trade system in 2000 that spanned 150 business units in more than 100 companies. Each unit was assigned a quota of emission permits. The company claims that the program cut greenhouse gas emissions by 10% below 1990 levels, 8 years ahead of schedule. BP was the first company to make a trade within the United Kingdom's Emission Trading System.

Royal Dutch Shell developed a pilot internal emissions trading system (2000-2002). The system allowed trading among group entities in Annex 1 countries. It covered 33 million metric tons of CO_2 equivalents from 22 separate sites. It subsequently established an Environmental Products Trading Business and entered the UK Emissions Trading Scheme. Shell Trading, with Nuon, executed the first trade in EU CO_2 allowances in 2003.

DuPont is a member of the Chicago Climate Exchange and International Emissions Trading Association. In 2002, it donated 120,000 tons of CO2e emission credits to Salt Lake City Organizing Committee allowing the Winter Olympics to be carbon neutral, by offsetting their emissions.[7]

With the growth in carbon emissions trading schemes, competition among financial firms and countries to gain a part of the growing carbon emissions trading industry has resulted. A July 6, 2007 *New York Times* article called carbon trading „the new big thing" and noted that the market was already worth $30 billion and could grow to $1 trillion within a decade.[8]

---

[7] Information from the Pew Climate Center, http://www.pewclimate.org.
[8] „In London's Financial World, Carbon Trading is the New Big Thing," *New York Times*, July 6, 2007.

The European Union's Carbon Emissions Trading System

In its effort to find cost effective ways to reduce emissions and fulfill its Kyoto Protocol obligations, the European Union (EU) decided to implement the world's first international carbon ETS. The system covers all 27 member states. Norway also cooperates with the program. The Directive mandated a system covering over 10,000 installations representing approximately 40% of CO_2 emissions in the power sector (facilities over 20MW), oil refining, cement, glass, ceramics, iron and steel, paper and pulp sectors. Each member state was required to set carbon allocation permits to companies operating in their own territories. Companies exceeding their CO_2 emission quotas would have to buy additional permits to cover their larger emissions. Companies making energy efficiency improvements or switching fuel sources could profit by selling off permits they no longer needed. In 2004, a Linking Directive was passed linking the joint implementation and clean development mechanisms of the Kyoto Protocol to the ETS. With this Directive companies can obtain credits by reducing emissions in developing countries and then using them within the ETS. The ETS was introduced in January 2005 and member states provided their allocation plans in May, which the Commission reviewed and approved after revisions.

At the national level, there were often lengthy domestic debates regarding emissions allocations and reduction targets. Within Germany, for example, there were harsh debates between the Environment and Economics ministers regarding what the reduction targets for the first (2005-2007) and second (2008-2012) periods should be.

The first year's assessment of the ETS suggested there were still many problems. Most importantly, it came to be realized that most countries had issued too many allowances for industry (at times at levels that were higher than actual emissions). This provoked a crash in the emerging carbon trading market. The price for one ton of CO_2 fell 63% from €30 to €11 from April 15 to May 15, 2005.[9] The market crashed again in April 2006 (to €8.3), when the actual figures for 2005 emissions came out, providing proof for the over-allocation, particularly in France and Germany.[10] The UK had the opposite problem. Having approved too few allowances, it appealed to the Commission for a change of plan,

---

[9] Spongenberg, Helena, „EU States Gave Too Many Pollution Permits, Say Environment Groups", *EUObserver.com*, May 15, 2005. http://euobserver.com/9/21594.
[10] Morrison, Kevin, „Lower Pollution in EU Sees CO2 Permits Fall 30%", *The Financial Times*, Aprili 27, 2006, p. 21.

but it was refused. This forced UK industry to buy 30 million tons of extra allowances.[11]

As is discussed more fully, in subsequent chapters, the European Commission began closely examining national allocation plans for the second phase of the ETS that began in 2008. The Commission, eager to see the system function well, took some measures to make sure that excessive allocations were not made again in the second phase of the program. The financial crisis and concomitant drop in oil prices have further challenged the system.

Carbon Emissions Trading in the United States

Somewhat ironically, given the historical role of the United States in developing the emissions trading concept, the United States Congress has failed to pass a carbon emissions trading scheme despite numerous bills being introduced into the House of Representatives and Senate. Thus, although the United States led in the development of emissions trading systems, other parts of the world have led in the development of carbon trading systems. The United Kingdom launched the world's first economy-wide emissions trading system in April 2002, and in 2005, the European Union launched the world's first international carbon emissions trading system. Other countries have announced plans to launch their own carbon emission trading systems (e.g. Australia) or are experimenting with voluntary schemes (e.g. Japan Voluntary Emissions Trading Scheme started in 2005).

Interestingly, although the US Congress has failed to introduce a carbon emissions trading system at the federal level, there interest in carbon trading at the regional level. The Conference of New England Governors and Eastern Canadian Premiers launched a regional greenhouse gas emissions trading scheme, most commonly simply referred to as REGGI (Regional Greenhouse Gas Initiative). California's Governor Arnold Schwarzenegger subsequently launched the Western Climate Initiative that links several western states and two Canadian provinces. These states have agreed to explore ways to work together to reduce greenhouse gas emissions, including with the planned launch of a carbon emissions trading scheme to be fully implemented in 2015. This scheme is expected to cover 90% of those states greenhouse gas emissions.[12]

---

[11]Euroactive, „Question Marks" http://www.euractiv.com/en/sustainability/question-marks-eu-co2-trading-scheme/article-155349?_print.

[12] The Western Climate Initiative, http://www.westernclimateinitiative.org/the-wci-cap-and-trade-program.

Conclusion

There are now close to three decades of experience with emission trading systems. Over time, there have been numerous important innovations. These include the development of banking (allowing firms to keep excess emission reduction credits for future use to encourage rapid reductions in emissions early on in a program), offset systems (permitting firms to offset their own emissions by making emission reductions elsewhere), multiple pollutant trading schemes, and the like.

Due to their general (or perceived) cost-effectiveness, emission trading systems are increasingly being used to address a wide variety of environmental problems. First used voluntarily to address local air pollution problems in the United States, emission trading systems have been used to control numerous kinds of pollutants (e.g. lead, SOx, NOx, chlorofluorocarbons, carbon dioxide, municipal waste,) and environmental quality and resource issues (e.g. water based nutrient trading, quota based fisheries management).

Initially a predominantly U.S. phenomenon, emission trading systems have spread to many parts of the world (e.g. New Zealand's fishing quota trading system, China's experimental sulfur dioxide trading system). They have also been incorporated into international regimes. The Montreal Protocol was the first international agreement that had emissions trading tied to it. Since this time, trading systems are becoming more common as implementation mechanisms for pollution control within international environmental agreements.

Yet, as subsequent chapters discuss, while there have been positive experiences with emissions trading schemes, there have been problems as well. Emissions trading schemes may lead to efficiency improvements, but they tend not to contribute to technological innovations. While they may be more efficient than some forms of regulation, they can also raise problems of equity and justice.

Searching for Meaning. Intersubjective Dimensions in Environmental Policies

Bettina Knothe

1 Introduction

Water, food, security, warmth and compassion for other beings are the most essential and existential demands on the environment and society for humans to survive. The organisation of these demands is the main driver of societal and political practice and follows specific traditional, historical, spiritual, cultural, socio-economic and very private rules. On the material side they depend on the availability of, and power over, land, natural resources and labour. On the immaterial side they depend on the deep sources of the embedded collective psychosocial experiences of past and present generations. Here, in the moderation of power, of the availability and value of material and immaterial goods, is the core factor for future sustainable living and production. This article discusses the thesis that the success of the internal implementation and external cooperation within European environmental policy is not *per se*, and solely, influenced by the choice of regulatory instruments but by the *quality* of the negotiations on the use of common pool resources.[1]

2 Blockades in natural resources management

Nature has always had a precarious double role in the ensuring of human sustentation: on the one hand, in the transfer, treatment and processing of natural resources for anthropogenic use, nature has been, and still is, a source of raw materials. On the other hand, nature serves as a sink for the disposal of waste products with more or less flexible options for regeneration. Since the availability of natural resources reached its limits in recent decades, the debate on the paradigm of sustainability has opened a highly complex and discursive field concerning mul-

[1] This article is part of the results of a research in cooperation with the Political Philosophy Group at the Università di Firenze, Italy, which was funded by a GARNET project mobility grant.

ti-level governance. This includes the description of global political, economic and social differences between North and South as well as the interwovenness between the discourses concerning ecology and social justice at the national and local levels.

This situation is confronted by the demand for an inter- and transdisciplinary examination of questions concerning equitable societal relationships and balanced opportunities to live a good life. It comprises demands for an equitable distribution of power, resources and rights among all human beings as well as reflections on modes of participation in the collaborative use of the limited pool of natural resources. Using the example of climate change Brunnengräber (2009) shows that during recent decades different discursive strategies have evolved in environmental policy debates. Regarding the consequences of anthropogenic influence on the climate, these discourses argue from different perspectives: some question the serious impact of the burning of fossil fuels in general and demand better proof of this in order to avoid cost-intensive measures. Others focus on the anthropogenic impacts and their consequences for the climate. Their discourse focuses on questions concerning the success of technological impact assessments, the mitigation or the reduction of emissions and measures for adaptation in terms of the implications, time patterns, locations and effects of climate change. A third approach attempts finding technical solutions to combat climate change. This debate strengthens strategies like sequestration (e.g. carbon capture and storage – CCS) as well as flexible compensation measures from the Kyoto Protocol such as Joint Implementation (JI, e.g. replacing a coal-fired power plant by more efficient combined heat and power generation) and clean development mechanisms (CDM, e.g. collecting and using credits from support for emission reduction projects in developing countries). Finally, a fourth approach aims at finding strategies for a so-called Green New Deal. This means supporting a change in the direction of equitable, fair and sustainable patterns of living and economic activities which comprise e.g. the use of renewable energies, new forms of labour and new industries. Protagonists of these strategies speak of a new social contract (Brunnengräber 2009).

But, societal and public spheres are not open to all kinds of socio-cultural expression. Instead, they consist of culturally specific and historically developed institutions, and of a certain social geography of space. Besides being arenas for the discursive development of opinions these public spheres are arenas for the development and production of social identities and constitutions (Fraser 1997). This influences and sometimes even hinders the successful implementation of new national and transnational environmental policy instruments. In the field of the management of European water resources this became obvious following the enactment of the European Water Framework Directive (WFD)[2] in 2000.

The intention of the WFD to influence national conditions with only little regulation, and to provide procedural methods instead, still demands detailed instructions which significantly minimise the national scope for interpretation on the one hand, and adaptations within existing national regulations on the other (Knill 2003). This framework increases the space for the diversity of the national modes of regulation, acknowledges the plurality of existing national regulation strategies and supports the maintenance of the scope for national arrangements within European demands. The context oriented patterns of the WFD define certain rules of procedure, but they do not define any substantial tasks with respect to expected outcomes. These regulative options are intended to facilitate an increasing flexibility in adapting to future developments and varying national ecological, economic, political and social constraints. But although the WFD is most influential for a national integrated water resources management, its implementation via network-based governance solutions has in reality proved particularly difficult for those Member States which have relied on hierarchical, sector-specific structures and regulatory instruments (Knill/Lenschow 2000, Rauschmayer et al. 2007).

Rauschmeyer et al. postulate that the challenges in the discourse on the governance of natural resources are mainly to be found in the fact that rhetorics of integrating the public and stakeholders, and of a science-based environmental policy, exist but are not being taken up in practice. On the contrary, meta-studies on participation and inclusion of scientific knowledge in the governance of natural resources show a lack of systematic and comparative studies, especially in specific fields of application, and particularly in studies addressing different fields. Although research on biodiversity and river basin management is increasingly expected to integrate science/policy interfaces in research design, this mostly happens without a systematic reflection on the aims and structures of such interfaces (Rauschmayer et al. 2007:4). Efforts towards a transparent governance of natural resources still remain in dichotomised patterns of apperceptions and actions which are built upon hegemonic relations to nature and which hinder the implementation of an equitable, environmentally adjusted and socially sound supply and care economy.

---

[2] Directive 2000/60/EC of the European Parliament and of the Council of 23 October 2000 establishing a framework for Community action in the field of water, Official Journal L 327 , 22/12/2000 P. 0001 – 0073.

3 Analytical dimensions of the societal relationship of humankind to nature

The paradigm of sustainable development was the answer to a crisis, which has meanwhile developed to become severely existentially threatening at an individual and societal level, e.g. in climate change conditions, the limited availability of resources, environmental catastrophes. Previous governance concepts often reduced complex problems to categories such as ‚global problems‘, ‚environmental problems‘ or ‚social problems‘ in order to make sustainable development feasible. They lacked the capability to explain the historical changes, the diversification of economic interests and concerns, and the movements of conflict lines in the meaning of time and space (Brunnengräber 2009).

Critical feminist and gender theories have pointed out that discourses on the environment and society cannot separate ‚nature‘ and ‚culture‘ into two independent entities with a more or less sustainable relationship. On the contrary, both are constructions of dichotomised patterns of recognition, awareness and appreciation by people in their environments as a whole (Haraway 1991). This thesis closely links two aspects: (a) symbolic attributions of everyday environments develop in the linkage of perceptions and attitudes within the interdependency of one's inner and outer world; (b) through the specific perceptions of the actors these attributions have a central function for the success of the organisation of different everyday spheres. From this perspective, socio-cultural patterns are diverse. These patterns are effective at the same time and in the same space but have different dynamics and qualities. They represent a very special and individual story-line within one's own environment, and they represent socio-ecological relationships which combine physical and symbolic dimensions of ecological, economic and social constraints.

This socio-ecological perspective explicitly considers relations rather than phenomena – relations between society and nature, both equally formed by social and ecological dimensions. In this perspective, a crisis always indicates the misbalance of those dependencies and relations within societal regulation. Environmental crisis is seen in close interdependency with social crisis and is interpreted as a ‚crisis of the societal relationship of humankind to nature‘ (Becker/Jahn 1989; Jahn/Wehling 1998; Becker/Jahn 2003). Here, nature and society are considered to be in a dialectical relationship to each other in which the assurance and functioning of social and economic existence is always linked to the assurance and functioning of the natural processes of life. The fulfilment of existential demands such as the supply of food, water, protection from extreme climate conditions, individual security and acknowledgement, work and production, sexuality and reproduction is the main driver, and economic production and gender relations are considered to be the main structuring factors.

Therefore, a precondition for new forms of political alliance which support the breakup of categorical and stereotypical dualities and functional attributions in environmental regulation may consist of the constructive acknowledgement of two factors: (a) the intense relation of the existence of human life to its ecological environment as an existential resource and (b) the intersecting factors which are responsible for identity-establishing and society formation. Those new forms are difficult to find, however. At the core, new alliances require the identification of new references in the sense of Hannah Arendt's *Zwischen*, which means in intervals and interspaces as public spaces for participation, personal and social performance, learning and emancipation. Only in these spaces do those *processes* of communication, understanding and negotiation become visible which form the basis for civil society's expression and commitment. These intermediate spheres are spaces for negotiation between – semiotically spoken – protagonist (*Akteur*) and complement/taker (*Aktant*) (Latour 1995, 2001). They are spaces between nature and culture, between institutions of the state and those of civil society, between people of different social origin and societal positions, between women and men, between experts and so-called laypersons. That which is normally hidden becomes visible here: the societal relationships of humans to nature are part of a routine, are produced and reproduced in everyday practices and become relevant in time and space. They are confirmed, promoted or hindered by specific forms of communication and mostly happen without much conscious reflection through implicit experiential knowledge and often under pressure of time and coordination. (Forschungsverbund Blockierter Wandel 2007).

In that sense, searching for strategic and methodological approaches to negotiating environmental problems with social dynamics and processes poses several questions. How can we talk about ‚environment' without being able to define more closely whose environment we address? Are we speaking about the environment of individuals, of societal spheres or of society as a whole? Do these different perspectives lead to a plurality of the subjective and objective concepts of the environment?

4   Political identity and shared meaningful intersubjective spaces

One factor all existing societies share is the need to distribute natural resources and to organise existential supply services. These are the connecting factors which run through all policy fields and intercultural relationships. For example, water management services are essential for constituting community life. Water is qualitatively and quantitatively (re)generated on the basis of geographical conditions which are locally and regionally specific. In contrast, on the level of regulation water resources are integrated into the global economy partly by transnational operators, as well as by factual and virtual trade. This makes communication about sound social and ecological water regulation difficult for political actors.

Societies have special strategies for regulating their relationships to nature materially and for symbolising them culturally. Existential relationships to nature exist and are profoundly anthropologically determined and culturally formed. On the one hand, they relate more or less to basic demands for human life and find their expression on the level of the personal/private, neighbourhood and community. On the other hand, due to globalisation processes, nation states with their specific civil society environment consisting of organisations, enterprises, media, non-profit organisations, citizens etc. are often extremely globalised in their professional and everyday practices. This is also true of the principles of the symbolic and semiotic reading of political regulation, such as norms for producing and consuming, lifestyles, patterns of awareness, culture etc. (Brand et al. 2007). In the space between the actual and potential possibilities the political field is a reservoir of unlimited varieties of negotiations and actions mediated through decision-making (Dryberg 1998). International political institutions and constructions are the result of certain patterns of behaviour and routines from strategies and projects as well as of the condensation of changing societal power relations, whereby national governments generally express national interests as a result of political compromises at the national level in order to be able to act autonomously in bilateral and multilateral negotiations. On the other hand, international institutions as well as international compromises and hegemonic projects influence national power relations and the re-formulation of national interests. In this interplay the condensed national power relations remain decisive for the characteristics of international institutions. At the same time, restrictions and incentives driven by international structures are effective at the national level and are materialised in nation-state institutions (Brand et al. 2007). Given that a political strategy is the vehicle of a systematic actualisation of constraints, the state serves as a terrain of social contention in order to build and sustain hegemonic positions. We also find this kind of systematic actualisation of societal constraints in the way resources are mobilised through discourses or cultural acceptance is achieved (Brand et al. 2007, Dryberg 1998). While policy is the practice of actualisation as such, the political process is the act of societal inscription.

As to the political structure of the EU, the distinction between political, social and cultural identities as well as the acknowledgement of the reflexive and discursive tension field of decision-making moves between a continuous negotiated common political identity and the diversity of cultures which are struggling with the (ir)reversible break from what used to keep them together. The interpretation of „making and implementing common decisions in high political questions and bearing the consequences that derive from them" (Cerutti 2005) implies a process-oriented notion of identity-establishment and is linked to the search for fostering communication regarding certain policy fields and reflexivity within intercultural relations. The question is how a political body like the EU can deal with resistance which mainly evolves at the level of nation states and which has secondary effects at the transnational level.

According to Cerutti (2008), legitimisation is still an essential condition for institutions and policies at the level of the nation states. This national political manifestation builds on specific national cultures and traditions which still make the development of a European identity difficult. As to political identity, Cerutti mainly follows a concept which is of a reflexive character and which is based on a form of inclusion, which reconciles the universalistic values on which the Union builds with the particularistic features of the European community by keeping the configuration open to these values. Political identity, in this perspective, is created by a set of political and social values and principles in which members of a polity recognise themselves as 'we'. But even more important than this set (identity) is the process (self-identification through self-recognition), „by which the people recognize themselves as belonging together because they come to share, but also modify and reinterpret those values and principles which are the framework within which they pursue their interests and goals". (Cerutti 2008:7).

Looking more closely at the role of individuals in political decision-making, people are influenced by an antagonistic tension of cognitive and normative frameworks of reference. On the one hand, identity building and cognition produce the concepts through which people conceive of themselves, the world and others. On the other hand, „identities are associated to socially binding norms according to which individuals evaluate the 'appropriateness' of an action [...]." (Lucarelli 2008:27). Lucarelli (2008) regards political identity „as a construction that is not and cannot be derived *directly* from a common culture, as 'the set of social and political values and principles that we recognize as ours, or in the sharing of which we feel like 'us', like a political group or entity'." (Ibid: 28) On the contrary, such values and principles need to be interpreted and identity, in consequence, is the „shared interpretation of a set of core values." (Ibid:28) Policy is both a context in which interpreted values can be observed at work, and an intervening variable in the process of identity-establishment. *Vice versa*, identity is an essential moment for the political process. Political entities receive legitimacy and acceptance because citizens are aware of certain values and principles and are willing to share (Cerutti 2001) and express them in their civil and every-day practices.

Everyday practices develop in various forms of negotiation processes concerning concrete issues and problems. These practices prove their suitability in specific civil contexts. They create transition paths between hegemonic and non-hegemonic spheres, become common through negotiation and are (re)produced in everyday practices. In this reflexive process everyday practices are always problem-oriented. They become spatially relevant through communicative support and confirmation or rejection. Here potentials develop such as personal commitment, responsibilities, relationships, interests, motivations, civilian resistance and originality, which often remain invisible in official political practice and are neglected in their productivity and commitment. These processes are free neither of contradictions and antagonisms nor of conflicts. But, especially due to this fact, in these intermediate spaces new interrelationships can develop and people can struggle for new situated contexts and meanings (Haraway 1995, Forschungsverbund Blockierter Wandel 2007). Spaces, locations and identities are negotiated on the basis of individual and joint sensitive and reflexive experiences (Massey 1994). They reveal the reservoir of interpretations which is available to the actors of the relevant societies as well as the character of individual commitment as a chance to link personal with social interest (Fraser 1997). These processes happen in symbolic spaces. The neurophysiologist Vittorio Gallese (2003) speaks of a ,shared meaningful intersubjective space' as a common interpersonal sphere of individual meaning and social seriousness. The psychoanalyst Jessica Benjamin calls this principle ,intersubjectivity' (Benjamin 1996, 1999, 2002). She regards the feeling of sharing presence with others and enjoying their subjectivity as enrichment and an important moment for abandoning one's own feeling of grandiosity for a shared feeling of empathy (Benjamin 1996).

5   Situated meanings of societal symbolisations

Probably a lot of people are familiar with certain occasions of changing cultural systems, surroundings and rationalities, e.g. when they are tourists, or when talking to colleagues at conferences or workshops, or when participating in further education activities. These occasions all have in common that the stay in the new systems is supported and safeguarded by the ordered, constant infrastructure at home. The family, friends, and opportunities (hopefully) are there to guarantee the existential basis of living. As a visitor a person can decide in a way how deeply he or she would like to integrate him- or herself into the new scene. The decision can be taken to leave this insecurity and return to a certain familiar terrain of safety or at least to a known situation of uncertainty and insecurity. At the moment a person decides, or is forced, to leave his or her own system, e.g. by migration due to poverty or oppression, and to live in new surroundings, in a new region, country, continent and culture, this situation changes completely. Now the entire personal existence has to be sustained in a surrounding of relationships, cultural habits, spiritual rites, forms of communication and attitudes towards life which is new, unknown and not embedded in the individual psychophysical constitution. Activities in the public and private spheres, e.g. the search for contacts and relationships, the establishment of work cooperation, the taking on of social responsibility, as well as the capability of securing one's own basic requirements for living, suddenly become extremely sensitive fields in which each individual reacts in her or his very personal way by asking her- or himself critical questions and drawing conclusions from concrete experiences. New experiences are more or less unconsciously evaluated against one's own biographically embedded and learned comprehension of, for example, respect and honesty as well as the feeling of, and strategies to deal with, reliability and friendship. In this wide range feelings may occur about what is personally considered to be constructive or offensive, respect or disregard, powerfulness or abuse etc. These feelings are closely connected with the capacity to cope with feeling oneself to be in contact with others, lonely and/or disconnected within the new surroundings. Current neuropsychological research emphasises a strong correlation between migration and stress, especially under socially isolated and uncertain constraints which result to a certain extent in severe mental disorders and psychological illness (Cantor-Graae, E., Selten, J.-P. 2005; González, H. M., Haan, M. N., Hinton, L. 2001; Karlsen, S., Nazroo, J. Y. 2002; Saraiva Leão, T. et al.2006; Van Os, J. 2003). In contact with a new societal organisation of life, personal *and* societal abilities determine whether a person is able to react to uncertainty with mutual curiosity and the passion to explore things, or with blockades and the fear to act.

This introduction shall strengthen two aspects: (a) cultural and traditional habits for comprehending, appreciating and organising existential supply and care requirements are deeply embedded in personal identity; (b) his or her 'equipment' with economic, cultural and social capital is the prerequisite for an individual person to be active and to organise her- or himself in a complex public space. This capacity also depends on the ability of a society to establish an equal, fair and inclusive environment. The quality of inclusion is mainly influenced by dimensions which are considered central factors influencing the gravity of oppression, such as gender, class and race. Since the 1990s the exploration of the interdependencies of these categories has increasingly moved into the focus of interest of gender, equality and migration research. The general assumption is that categories like sex, gender, ethnic origin, class, age and physical condition determine identity and subjectivation. They interweave or ‚intersect' one another. On the other hand, identity develops in structures and conditions which are also intersectional at a societal level. So, in contrast to simply adding up the impacts of several patterns to a syndrome of oppression, the concept of 'intersectionality' means acknowledging that the dichotomised dimensions of inequality are interwoven and mutually influence each other (cf. Degele/Winker 2007, Winker/Degele 2009). Based on a longer tradition within the feminist movement of the critical observation of the global sisterhood between white and black women, the concept of intersectionality leads to a comprehensive theoretical and empirical analysis of the importance these different categories have for inequality and oppression. This indicates that intersectional analysis works on the basis of dominance (*Herrschaft*) approaches. Following Harding (1991) and Degele/Winker (2007) intersectionality as a multi-level framework for the analysis of hegemonic power relationships acknowledges three levels of representation: (1) identity structures, (2) societal structures and (3) symbolic representations. Although they mainly refer to the interdependencies between the social categories gender, class and race, the categories sex, age, (dis)ability, religion and nationality are generally compatible enough to be integrated (Degele/Winker 2007, Winker/Degele 2009). Intersectional research on inequality among other things, therefore, treats the question of how specific social categories (see above) affect societal discrimination and privilege. By doing so, the different categories are not analysed in parallel but treated as interwoven, intersecting with one another, and – according to the concrete context – influencing one another. Intersectional analysis in this context means examining the structures, including institutional structures and symbolic contexts, in which social practices are embedded, and how they create, modify and differentiate identities and find their expression in the everyday activities of different actors. At the level of identity this enables people to ask for alternative possibilities and arrangements for commitment. This means, first, that societal change can be better understood and, secondly, that theoretically verified factors for political action can be derived.

6   Precaution as a core value for intersecting spheres of supply services

The maintenance of the (un)equal distribution of resources through practices of inclusion and exclusion can be examined on all these intersecting levels of representation. This presents the opportunity to extract hegemonic norms and stereotypes which are reproduced in everyday life in order to contribute to one's own subjectivation and to support the existing conditions for power and dominance. Central in this case is the efficacy of discourses which – on the basis of mostly implicit knowledge – are the constantly repeated and reproduced practices of language, communication and experience. For example, the phrase ‚human failure' expresses the fact that routines which were very successful in everyday life for a certain time suddenly, in the case of a disaster or catastrophe, become visible and conscious. While reflecting the course of the event they now turn out to be no longer appropriate. This situation demands completely new solutions instead of keeping to familiar habits in the sense of having-done-this-since-ever.

Combining the socio-ecological with the intersectional perspective in the shaping of political decision-making processes for natural resources management and for a fair and equal functioning of supply infrastructure, we may assume that the specific demands of rural and urban households have to be the starting-point for the negotiation of differentiated infrastructure models. Citizens regenerate and differentiate knowledge and competency for societal issues in the reproductive interdependencies of their everyday experiences. This knowledge is highly relevant and useful for participation in supply infrastructure because it contributes embedded, culturally »situated« knowledge and expertise in societal services (Knothe 2007). This aspect is closely linked with the difficult issue of how to handle the unknown and unpredictable in a constructive goal-oriented negotiation process.

Strong sustainability from a socio-ecological perspective would on the one hand allow for the uncertainty of future preferences of the user system. On the other hand, flexibility for differentiated technologies, e.g. for water management to withdraw and produce freshwater as well as for the collection and treatment of wastewater, would acknowledge the multifunctional character of, and demands on, the regeneration of ecological systems. Establishing regulatory terms to deal with the unpredictable means that precaution for today as well as options for future generations are taken into consideration. Connecting strategies of strong sustainability with intersectional perspectives, natural resources management also means acknowledging hegemonial relations between the professional administrative and economic institutions and committed actors and groups within civil society. This requires the creation of a shared meaningful intersubjective sphere for all the actors involved in order

a.   to open spaces for developing sensitivity towards societal processes which create identity and towards a comprehension of the relation between the ‚inner' and the ‚outer' environment;
b.   to discuss and transform conditions which influence the dynamics of inclusion and exclusion of societal groups within decision-making processes;

c.   to examine the often unequal relationship between different categories of knowledge (professional – everyday, theory – practice);

d.   to create open but safe spaces in order to develop a sensitivity for immanent processes of ‚othering‘ and to make categorisations of ‚the other‘ visible and therefore accessible for discussion and transformation.

‚Care‘, ‚responsibility‘ and ‚precaution‘ are considered to be elements connecting the productive and reproductive spheres, the conservation *and* formation in the production of goods and services in the shaping of social systems (Knothe 2007). They are intersectional under socio-cultural, ethnical and historical perspectives and related to space. Spatial identities, evolved during various biographical and societal histories and accompanied by specific practices of their constitutions, are both material and discursive (Massey 2004). Assuming that personal and cultural identity is bound to place, it not only matters how both ‚place‘ and ‚identity‘ are conceptualised internally but also how this conceptualisation is legitimised externally and how these identities and the social geography of their relations can be connected with the performance of political responsibility and legitimacy. Negotiation processes will then be successful if it is acknowledged that new results of participation and governance can emerge from insecure and performative action in fluent processes. This needs space for various knowledge practices, repetitions and a step-by-step approach towards decisions, creativity, combination and inventive solutions, everyday social practices of the know-how, the space and the comprehension of those realities which are created by situated action. In this sense, everyday practices offer every potential for the placement and embodiment of knowledge and experience instead of solely demand for abstract insights, which are anonymous and therefore without links to direct responsibility (Haraway 1995). This kind of negotiation enables the upgrading of knowledge assets, which are attained and stored in everyday life through practical, intuitive, incorporated biographical knowledge.

Political practice should allow spaces for action by participants, actors, partners and projects, so that they can leave their safe positions and enter into an open discussion of conflict. This requires a situation in which the participants are able to reflect on their own situated character and their contextualised status of an observer against one-sided models of explanation. In doing so, opportunities may occur for a new quality of situated adequateness in which chances for new unexpected coalitions can be identified, taken up and put into practice. Institutional acknowledgement and the stabilisation of already existing experimental spaces for negotiation through a policy which stabilises continuous exchange is needed. This kind of intermediate space for sustainable development requires a new framework. It may be a kind of problem-oriented alliance which collects diverse information and accompanies experimental phases. It builds upon an empathic basis of the patient and mutual comprehension of different interests and possibilities for problem-solving, (un)realistic alternatives and various possibilities of interpretation of potential side effects and interplays.

Immediacy and pressure are often instruments of hegemonic practices of power. They hinder a constructive reflection of routines and the discussion of inherent necessities and assumed matters, of course. Plurality in negotiation processes therefore needs time and the relief of pressure for action. Time is an elementary precondition in order to attain insight into complex processes, to develop unprejudiced and well-founded problem solutions far from urgencies, time pressure and withdrawal to the assumed common welfare. Under this perspective, patience and prudence become essential competencies of continuous accompaniment. This offers the opportunity to create constellations of actors in which results are assured at all levels of political regulation and experienced knowledge is effectively stabilised. For the official administration, prudence in dealing with negotiating processes means developing sensitivity towards differentiated demands as well as the proactive involvement of citizens. The diversity of activities in interspaces leads to the fact that political institutions have to search for and acknowledge the invisible, to wait for the unexpected and to value the devalued. One task of administrative institutions has to be to create possibilities for actors to reflect their experiences and knowledge and connect with one another. Institutions need to bear the unexpected which develops from the situation that decisions may augment the number of possible futures and future options (Kropp 2002). Being able to cope with the unexpected at an institutional and civil society level is the precondition for demystifying participation. It empowers and enables the (fore)seeing of actions and competencies beyond not yet agreed standards. This means acknowledging the inevitability of not knowing and insecurity. By doing so, the institutional hegemony of experts and strict administrative and technocratic standpoints will be limited in favour of a democratic processing of knowledge inputs in learning processes, which create sensitivity for the plurality of perspectives and alternatives in civil societies.

References

Becker, E.;Jahn, T. (2003): Societal Relations to Nature. Outline of a Critical Theory in the ecological crisis. In: Böhme, Gernot; Manzei, Alexandra (ed.): Kritische Theorie der Technik und der Natur. Munich: pp. 91-112.

Becker, E.; Jahn, T. (1989): Soziale Ökologie als Krisenwissenschaft. Sozial-ökologische Arbeitspapiere Nr. 1. Frankfurt am Main: Verlag für interkulturelle Kommunikation.

Benjamin, J. (1996): Phantasie und Geschlecht. Psychoanalytische Studien über Idealisierung, Anerkennung und Differenz. Frankfurt am Main: Fischer Verlag.

Benjamin, J. (1999): Die Fesseln der Liebe. Psychoanalyse, Feminismus und das Problem der Macht. Frankfurt am Main: Fischer Verlag.

Benjamin, J. (2002): Der Schatten des Anderen. Intersubjektivität, Gender, Psychoanalyse. Frankfurt am Main, Basel: Verlag Stroemfeld/Nexus.

Brand, U.; Görg, C.; Wissen, M. (2007): Verdichtungen zweiter Ordnung. Die Internationalisierung des Staates aus einer neo-poulantzianischen Perspektive. In: Prokla 37/2007(2): pp. 217-234.

Brunnengräber, A. (2009): Die politische Ökonomie des Klimawandels. Munich: oekom Verlag.

Cantor-Graae, E.; Selten, J.-P. (2005): Schizophrenia and Migration: a Meta-Analysis and Review. In: American Journal of Psychiatry 162: pp. 12–24.

Cerutti, F. (2005): Constitution and Political Identity in Europe. In: Liebert, U. (ed.): Postnational Constitutionalisation in the Enlarged Europe: Foundations, Procedures, Prospects. Baden – Baden: Nomos. Available at http://www.sifp.it/articoli.php?id Tem=3&idMess=91, accessed April, 4th 2008.

Cerutti, F. (2001): Towards a political identity of the Europeans – An introduction. In: Cerutti, F.; Rudolph, E. (eds.): A soul for Europe, Vol. 1, A Reader, Edition Peeters, Leuven: pp. 1-31.

Cerutti, F. (2008): Why political identity and legitimacy matter in the European Union. In: Cerutti, Furio; Lucarelli, Sonia (eds.): The search for a European Identity. Values, policies and legitimacy of the European Union. London, New York: Routledge/ GARNET series: Europe in the World: pp. 3-22.

Cerutti, F.; Lucarelli, S. (eds.) (2008): The search for a European Identity. Values, policies and legitimacy of the European Union. London, New York: Routledge/ GARNET series: Europe in the World.

Degele, N.; Winkler, G. (2007): Intersektionalität als Mehrebenenanalyse. Available at http://www.tu-harburg.de/agentec/winker/pdf/Intersektionalitaet_Mehrebenen.pdf, accessed March, 3rd 2008.

Dryberg, T. B. (1998): Diskursanalyse als postmoderne politische Theorie. In: Butler, J.; Critchley, S.; Laclau, E.; Žižeck, S.: Das Undarstellbare der Politik: zur Hegemonietheorie Ernesto Laclaus. Vienna: Verlag Turia+Kant: pp. 23-51.

Forschungsverbund Blockierter Wandel (2007): Blockierter Wandel? Denk- und Handlungsräume für eine nachhaltige Regionalentwicklung. Munich: oekom Verlag.

Fraser, N. (1997): Die halbierte Gerechtigkeit. Frankfurt am Main: Suhrkamp.

Gallese, V. (2003): The roots of empathy: The shared manyfold hypothesis and the neural basis of intersubjectivity. Psychopathology 36, 2003: pp.171-180.

González, H. M.; Haan, M. N.; Hinton, L. (2001): Acculturation and the Prevalence of Depression in Older Mexican Americans: Baseline Results of the Sacramento Area Latino Study of Aging. In: Journal of the American Geriatrics Society 49, 2001: pp. 948–953.

Haraway, D. J. (1991); Simians, Cyborgs, and Women. The Reinvention of Nature. New York: Routledge.

Haraway, D. J. (1995); Die Neuerfindung der Natur. Primaten, Cyborgs und Frauen (edited and introduced by C. Hammer & I. Stießl), Frankfurt am Main/New York: Verlag Campus.

Jahn, T.; Wehling, P. (1998): Gesellschaftliche Naturverhältnisse – Konturen eines theoretischen Konzepts. In: Brand, K.-W. (ed.): Soziologie und Natur. Theoretische Perspektiven. Opladen: Leske+Budrich: pp. 75-98.

Karlsen, S.; Nazroo, J. Y. (2002): Agency and Structure: the Impact of Ethnic Identity and Racism on the Health of Ethnic Minority People. In: Sociology of Health and Illness 24(1), 2002: pp. 1–20.

Knill, C. (2003): Europäische Umweltpolitik: Steuerungsprobleme und Regulierungsmuster im Mehrebenensystem. Opladen.

Knill, C.; Lenschow, A. (2000): On deficient implementation and deficient theories: The need for an institutional perspective in implementation research. Implementing EU Environmental Policy. New Directions and Old Problems. Manchester, New York: Manchester University Press.

Knothe, B. (2007): Elements of care in water management. Gender dimensions for sustainable water use. In: Lozán, J. L. et al. (ed.): Global change: Enough Water for all? Hamburg: GEO-Verlag: pp. 322-324.

Kropp, C. (2002): ‚Natur' soziologische Konzepte, politische Konsequenzen. Opladen: Verlag Leske+Budrich.

Latour, B. (2001): Das Parlament der Dinge. Für eine politische Ökologie. Frankfurt am Main: Suhrkamp.

Latour, B. (1995): Wir sind nie modern gewesen. Versuch einer symmetrischen Anthropologie. Berlin: Akademie Verlag.

Lucarelli, S. (2008): European political identity, foreign policy and the Others' image. In: Cerutti, F.; Lucarelli, S.:The search for a European Identity. Values, policies and legitimacy of the European Union. London/New York: Routledge/GARNET series: Europe in the World: pp. 23-42.

Massey, D. (2004): Geographies of Responsibility. Geografiska Annaler, Series B: Human Geography 2004/86/01: pp. 5–18.

Rauschmayer, F.; Wittmer, H.; Paavola, J. (2007): Multi-level Governance of Natural Resources: Tools and Processes for water and Biodiversity Governance in Europe – A European Research and Training Network. Available at http://www.governat.eu/ library, accessed February, 17th 2011.

Saraiva Leão, T. et al. (2006): Incidence of Schizophrenia or Other Psychoses in First and Second-Generation Immigrants. In: The Journal of Nervous and Mental Disease 194, 2006: pp. 27–33.

Van Os, J. (2003): Discrimination and Delusional Ideation. In: British Journal of Psychiatry 182, 2003: pp. 71–76.

Winker, G.; Degele, N. (2009): Intersektionalität. Zur Analyse sozialer Ungleichheiten. transcript Verlag.

On the Way to the Future – Renewable Energies

Lutz Mez / Achim Brunnengräber

In order to mitigate and stop the threatening climate change, the existing fossil and nuclear energy system must be replaced by a lasting energy system within the next few decades. Following IPCC recommendations, greenhouse gas emissions must be more than halved worldwide until 2050 in relation to the emission level of 1990. This is forced not only by the scarcity of fossil resources and the resulting conflicts, but also by the progressive social and ecological threat as well as the economic risks associated with climate change. Renewable energies (RE) play a key role.[1] They contribute to world peace, to the protection of the world climate and the environment, create new jobs and vocational fields and democratise through their decentralised energy supply, which today is strongly concentrated on only a few players.[2] In addition, renewable energies serve the preservation of resources, help fighting poverty in the developing countries, and promise the lowering of imported fossil and nuclear energy sources to Germany, the European Union, and to all other countries dependent on fossil energies. On a long-term basis, they thus contribute to energy supply security and to human security.

In relation to these hopes, which are often linked with renewable energies, the reality looks completely different. The Kyoto Protocol as the central international answer to climate change is hardly able to cause or even activate the change of the energy system. However, as we know today (see Nell, Semmler and Rezai in this book), the flexible instruments of the Kyoto Protocol are only insufficiently suitable for it. Emission trading has hardly effected the promotion of renewable energies, while the German Renewable Energy Sources Law (EEG) is aimed at technology and industrial policy and not on the reduction of greenhouse gases.[3] Both instruments must be regarded as „two separated strands of the

---

[1] Renewable energy sources are wind, hydropower, biomass, solar and geothermal energy.
[2] In contrary to the targeted liberalisation of the European electricity and gas market we can observe higher concentration and giant mergers: e.g. E.ON and Ruhrgas in Germany or French GdF and French-Belgian Suez.
[3] According to an analysis of DIW, DLR, ZSW and IZES (2008) for the Federal Ministry of the Environment, emissions trading and EEG are in conflict. Industry will use all emission rights for the

energy and climate policy" (for details see Fischedick 2008:103). Existing instruments must be reformed and strengthened, and new instruments must be developed in order to establish a lasting and efficient transformation and usage of energy sources and electricity. The target is to replace fossil energies (coal, oil, and gas) and nuclear power by an environmentally friendly renewable energy system on a mid-term and long-term basis, in order to actually and clearly decrease greenhouse gas emissions. Further technology and the transfer of know-how to newly industrialising and developing countries has to be intensified, so that the possibilities of the development of renewable energies can be used globally. Furthermore, renewable energies also have the potential to develop the self-sufficiency of developing countries and to reduce their dependence on imported energy.

The article discusses the possibilities but also the limitations of a radical transformation of the energy system, which aims at a complete substitution of fossil energies in mid-term and long-term perspectives. Firstly (1), the new limits to growth are demonstrated and (2) it is argued that nuclear power does not make a contribution to climate protection. Then (3) the 2020 targets of the European Union (EU) for renewable energies are presented and (4) different promotional instruments used at present in Europe and worldwide for the development of renewable energies are analysed. Furthermore (5), the necessity for international networking and the institutionalisation of players who spread and support the knowledge of different renewable energy sources and the transfer of technology is depicted. In order to promote the worldwide transfer of renewable energy technology and the increase of the energy efficiency, the „International Agency for Renewable Energies" (IRENA) is introduced as an important step. In the concluding part (6), the democratic relevance of renewable energies from national perspectives and an additional North-South perspective is treated.

1   Persistence of the conventional energy system

Since the first report of the Club of Rome on "The Limits to Growth" (Meadows et al. 1972), the exhaustion of fossil energy sources, in particular oil, coal, and gas, is discussed. The purpose of the report was not to make specific predictions but to explore how exponential growth interacts with finite resources. Since this publication, public consciousness over the finiteness of fossil resources was never lost. When Graham Turner (2008) examined the past thirty years of reality

---

period 2008-2012 of 453 million tons CO_2 per year. The expansion of RE will reduce the price of certificates and lower the steering effect of emissions trading.

on the basis of the predictions made in 1972, he found that changes in industrial production, food production, and pollution are all in line with the predictions of an economic and societal collapse in the 21st century. In the course of rising energy prices, the debate about the question of scarceness of energy resources and how energy supply and security can be ensured in the future resurfaced. In the age of climate change, however, the ecological limits must be defined differently today than nearly 40 years ago. The increasing water shortage, the expansion of deserts, and the loss of biodiversity do not represent classical environmental problems but interacts closely with a number of social problems: the worsening of the hunger crisis and pauperisation, the increase of environmental migration, and the conflicts about scarce resources, such as soil and water. Worldwide dramatic conflicts already revolve around agricultural farmland and fresh water supply. The socio-ecological and socio-economic crises resulting from it must be urgently counteracted. Not only the finiteness of resources, which will be reached in a few decades, is the main problem of mankind but also the effects of fossil energy consumption and the comprehensive and irreversible destruction of the ecological bases of life. Due to the social, ecological, and economic subsequent effects of climate change, the promotion of fossil energies has to be stopped – and long before the last drop of oil from the earth's crust is pressed.

For social and ecological reasons, the energy industry thus already stands at a crossroads. Instead of large central power stations – „the cathedrals of electricity" – distributed generation technologies, e.g. small-scale CHP plants, with completely different social relations between energy producers and consumers have a historical chance (Lönnroth 1996). The fossil age could soon be replaced by the solar age. However, the political and cultural frame conditions still have to be created (for details on the cultural obstacles connected to the change of the energy system see Amery/Scheer 2001). Politico-economically, however, the concentration of power in the energy sector is the biggest problem. Over the course of the 20th century, the electricity industry has become a major player in the energy policy process in almost every country of the world, be it in the West, East, North or South. It often acts as a state within the state with a power base that could not be diminished but instead broadened over time. In its history, neither changing institutional conditions nor free market ideologies like free trade or the political system change after the end of the East-West confrontation were able to change this industry. In its origins, it revolves strongly around technologically driven developments, which favour the emergence of powerful structures within the grid-bound energy industry. The electricity supply industry serves as a good example for these kinds of processes.

In the industrialised countries, the development of electro-technology began with a symbiosis of utilities and electrical industry. Firstly, companies such as

Siemens, Lahmeyer and AEG established subsidiary companies in order to be able to sell turbines and generators as well as other parts for power plants. Electricity was supplied first for lightening and power machines. The electrical industry, in turn, produced light bulbs, electric motors, and other electrical appliances, which secured and expanded electricity sales. The symbiotic relationship between the electric industry and the electrical appliance industry partly dissolved over the course of time. From the year 1900, the first utilities emancipated themselves from their parent companies (for example RWE, see Mez/Osnowski 1996). After the Second World War, the electric industry tried again to take over certain functions of the utilities and offered turnkey nuclear power stations. That led to the fact that the design and construction departments of the utilities were dissolved.

In the period between the Second World War and the first oil price crisis in 1973, the energy policies of all industrialised countries followed the same patterns of development. The so-called „energy syndrome" (Lindberg 1977) arose at the same time in the East and the West; it was the result of simultaneously arising symptoms, which led to a system failure. Two of the symptoms are the facts that energy production and consumption required a constantly growing power supply and that the interests of energy producers determined national energy policy. After more than two decades, a shift of paradigm in energy policy is on the agenda in industrial as well as in developing countries for several reasons.

Most prominently feature supply uncertainties around fossil energies and the ecological problem pressure due to which the industry and political decision makers are confronted by the public and the environmental movement (see Altvater in this book). During this process, environmental and interests of nature protection are mobilised, which can influence economic and political decisions. As a consequence of the oil price crises, demands arose for an economical use of energy. Landscape protection put brown coal open-cast mining or power generation in large hydro-electric power plants on the agenda. And due to the most recent events in Fukushima, radiation protection is a hot topic on the political agenda of all Western industrialised countries. Since the late 1980s, climate protection requires new aspects of a rational use of energy in order to decrease CO_2 emissions.

But despite many successes, growing public awareness and new limits to growth, which go hand in hand with climate change, loss of biodiversity, desertification and freshwater scarcity, and with the limited sink capacity of the earth for wealth garbage, the chosen fossilistic energy path still was not abandoned (for „the new limits of growth", also see Meadows et al. 1993). The development of renewable energies is painfully slow: Since 1990, the average annual growth of renewable energy production is only 1.8%. This corresponds roughly to the rate of growth of the world's primary energy supply of fossil fuels. The share of renewables in global primary energy consumption in the year 2008 was 13.1%, which amounts to 18.5% of final energy consumption.[4] But the share of modern solar, wind, and tidal energy to the renewable energies is still marginal. Within the worldwide generation of electricity in 2008, it only amounts to 2.8% and 3.1% of the global final energy consumption (for details, IEA 2010). The era of large power plant units and condensing power plants would be finished by small, economical and distributed systems as well as through renewable energy technologies. The development of a sustainable energy system could re-strengthen the link between the electric industry and utilities, and create innovations and new electrical engineering. This development can cause a potential of growth for renewable energies and could provide a faster access.

2 Climate change and nuclear power plants

After almost 25 years, the disaster of Chernobyl appeared to be long forgotten. The debate about a renaissance of nuclear power – after "decades of stagnation" (Rosenkranz 2006: 36) – showed that the catastrophe was barely remembered and that nuclear power was instead seen as the energy source to protect against climate change and global warming. According to the "Energy Technology Perspectives", published by the International Energy Agency in June 2008 for the G8 summit, in addition to other measures, 32 new nuclear power plants were to be built per year to reduce 50% of harmful greenhouse gases by 2050 (IEA 2008). The most recent events in Fukushima, Japan, however, brought forth a drastic turn in the debate. Following the March 11, 2011 earthquake and the tsunami that resulted from it, the nuclear power plant Fukushima Daiichi was heavily damaged which caused a core melt-down in several reactors and contaminated a wide area surrounding the plant. The severe damages in the Fukushima reactors were rated as major accident at level 7, this is the same level of the INES scale as the Chernobyl disaster was rated. That such an accident could happen in a high tech country like Japan generated a worldwide impact.

Apart from the security debate that arises, this only shows that upgrading nuclear power as a technology for climate protection is a popular fallacy: There are currently 437 nuclear power plants (NPP) worldwide, which operate in 30 countries (PRIS 2011). However, just six countries – the U.S., France, Japan, Germany, South Korea and Russia – cover over 70% of globally generated nu-

---

[4] The high share is based mainly on the use of biomass, particularly in the developing countries. Furthermore large hydro-electric power plants still generate the main load of electricity.

clear electricity. The contribution of nuclear power to the world energy supply is currently at a relatively low level of 5.8% – only half the share of renewable energies. In 2008, electricity generation accounted for 17.2% of global final energy consumption and only 13.5% of those were produced by nuclear power stations. Thus, the share of nuclear electricity is less than 2.3% of total global final energy consumption (Mez, Schneider & Thomas 2009).

For several reasons, the share of nuclear power cannot increase significantly in the mid-term or long-term perspective. About six times as many nuclear power plants would be needed to be installed around the world than exist today, which would be an order of 2.100 large NPPs with a capacity of 1.000 megawatts. In the study, the IEA assumed investments of 45 trillion U.S. dollars necessary to stop climate change. Besides the construction of 1.400 nuclear power plants, a massive expansion in wind turbines has to be financed (IEA 2008). The German energy expert Klaus Traube (2005) gives cause to serious concern that new nuclear power plants would have to be built in all regions of the world. The danger of terrorist attacks on NPPs would rather increase than decrease. Finally, the world's uranium reserves are limited and currently about 70.000 tonnes of uranium per year are required to operate the existing nuclear power plants. A massive expansion of nuclear power would need an additional 260.000 tons of uranium per year. As a consequence, the currently known reserves of uranium would be spent not after 70 years, as is currently assumed, but rather after only 18 years.

The carbon footprint of nuclear power plants also disfavours the nuclear expansion. Green house gas emissions cannot be calculated solely on the basis of actual operation but have to take the entire production cycle into account. Then it quickly becomes clear that nuclear power plants are no CO_2-free production systems, as presented in the advertisements of the operating companies. Already, nuclear power emits up to one third of CO_2 as traditional gas-fired power plants. Production-related CO_2 emissions of nuclear energy – depending on where the raw uranium and the fuel rods are produced – add up to 126 grams of CO_2 per kilowatt-hour. The Öko-Institut (Fritsche 2007: 7) has calculated the carbon footprint for a typical nuclear power plant in Germany with 32 grams of CO_2 equivalents. By comparison, a natural gas cogeneration plant emits 49 grams, an import-coal CHP plant 622 grams and a traditional lignite power plant is responsible for 1,153 g/kWh CO_2 equivalents. Much better is the carbon footprint of biogas total energy units with minus 409 g/kWh and also the carbon footprint of wind power with only 24 g/kWh (all data see Fritsche 2007).

In addition, nuclear power plants are emitting other gases, which also contribute to climate change. Of all radioactive materials, the ionisation of the air with radioactive noble gas, krypton-85, a product of nuclear fission, is the most intense. Krypton-85 is generated at nuclear power plants and is the last released in reprocessing plants or during conditioning of nuclear waste. The concentration of krypton-85 in the atmosphere has strongly increased during the last decades by nuclear fission and reprocessing of nuclear fuel. In 1994, the concentration of krypton-85 on the Northern hemisphere was 1 Bq/m^3 and a little less in the Southern hemisphere. The natural concentration of krypton-85 is only 0.000 000 01 Bq/m^3 (Koller & Donderer 1994). But neither in the current climate policy debates nor in the Kyoto Protocol does krypton-85 play a role. Krypton changes the conductivity of the air and the air-electrical system of the atmosphere. The high expectations for nuclear power have therefore – from a few NPP operators and nuclear power advocates, however apart – given way to the insight that nuclear power is a phase-out model, because there are always new risks shown.[5] Unresolved problems remain: the final disposal of radioactive waste, the limited uranium reserves, and the fast breeder technology (FBRs should produce plutonium as new nuclear fuel and expand the range of uranium resources by factor 60) has hitherto failed.

3   Policy targets of the European Union

If not necessarily from a political-institutional point of view but definitely for their problem structure, climate and energy policy are two sides of the same coin because 70% of CO_2 emissions result from the production or transformation of energy (for "strategic selectivity" between climate and energy policy, see Brunnengräber 2007). If the EU reduces its greenhouse gas emissions by at least 20% by 2020, the consumption of fossil energies has to be reduced as well. Therefore, the European Commission's approach to interlock and combine climate and energy policy is absolutely right. The European double-track strategy intends to improve energy efficiency, to counteract the increasing dependency on imports, and to further extend the use of renewable energies. How this objective of reducing the consumption of fossil fuels while competition and growth orientation are to be implemented at the same time still remains unanswered in the strategy papers of the European Union.[6]

The hope lays mainly on the development of renewable energies, which – as already shown above – are not supported by the Emission Trading System. In

---

[5] To the dangers of NPPs, the risks for human security and repeatedly occuring accidents see in detail Schneider et al. 2007

[6] Even the latest European Council Conclusions of February 4, 2011 on energy, innovation etc. contain for energy policy only the broad formula of „better coordination of EU and Member States' activities".

January 2007, the EU Commission presented its energy and climate policy package and an ambitious policy on renewable energy and outlined a "roadmap". On this basis, the European Council, in March 2007, decided on binding targets for renewable energy development. By the year 2020, the share of renewable energy has to be increased to 20% of the EU's final energy consumption. This is almost a threefold increase of the present situation. By 2020, 10% of fuel consumption must come from renewable sources (European Commission 2008a) – in spite of all social conflicts due to the competition of foodstuff for hungry stomachs and biofuels for empty tanks (see Brunnengräber in this reader).

To implement these objectives, the EU Commission, in January 2008, tabled a new Directive on renewable energy. In December 2009, the directive was adopted by the European Council and the European Parliament. The directive requires every nation in the 27-member bloc to increase the share of renewable energy from 5.5% in 2005 to 20% by 2020. This target applies to total final energy consumption in the electricity, heating and cooling, and transport sectors.

The directive set interim targets in order to ensure steady progress towards the 2020 targets. 20% average between 2011 and 2012; 30% average between 2013 and 2014; 45% average between 2015 and 2016, and; 65% average between 2017 and 2018. EU countries are free to decide their own preferred „mix" of renewables, allowing them to take account of their different potentials. Under great time pressure they had to present national action plans (NAPs) to the European Commission by 30 June 2010, in which targets for the three sectors were defined and which states that progress reports shall be submitted every two years.

„The starting point, the renewable energy potential and the energy mix of each Member State vary. It is therefore necessary to translate the Community 20% target into individual targets for each Member State, with due regard to a fair and adequate allocation taking account of Member States' different starting points and potentials, including the existing level of energy from renewable sources and the energy mix. It is appropriate to do this by sharing the required total increase in the use of energy from renewable sources between Member States on the basis of an equal increase in each Member State's share weighted by their GDP, modulated to reflect their starting points, and by accounting in terms of gross final consumption of energy, with account being taken of Member States' past efforts with regard to the use of energy from renewable sources" (European Commission 2009).[7]

---

[7] Renewables play a very different role in the energy mix of the European Union member states momentarily: While for instance Sweden, according to EU commission data, in 2005 already had a share of 39.8%, the share in the United Kingdom was only 1.3% and in the Netherlands 2.4%. Ger-

It might be difficult to achieve the 10% target for biofuels since the second generation of biofuels would still need to be economically and ecologically sound and the Fuel Quality Directive should be amended accordingly. Also in the field of energy efficiency, a lot can be achieved. The car industry, so far, successfully defended itself against strict regulations of exhaust emissions or impact of road tax and is actively supported by the governments of Germany and France. Impetus for innovation, particularly in consumer goods – according to the Japanese model –, could derive from a European top runner program. It would help the most efficient technologies to penetrate the market much faster. The consumption of most efficient devices – combined with significant savings within five years – becomes the binding standard for the industry.[8] Also in the transport sector, potential efficiency gains could be realised quickly. But a variety of policy measures and politics need to be adopted in the Member Countries, which have not even been implemented to some extent so far. Germany, as the alleged leader in climate protection policy, has set ambitious targets for the reduction of greenhouse gas emissions. The necessary far-reaching measures at national level – as in the European Union – fail because of the orientation of governments and industrial players towards international competitiveness and growth.

4   Support instruments

Support schemes for renewable energy are developed for a fundamental transformation of the energy sector – and not primarily to reduce emissions, but to build up a powerful eco-industry. And of course the instruments shall support the introduction of new technologies, in research and development or at the stage of pilot and demonstration plants. Renewable energy sources are part of an energy mix to ensure energy security and to generate new export opportunities. Since the late 1980s, various support systems for developing renewable energy sources such as the feed-in tariff and the quota obligation model were developed and applied in the countries of the European Union. The Commission has not succeeded in developing a harmonised system for the promotion of renewable energies. Today, a variety of support instruments exist for renewable energy in the electricity, the heating, and the transportation sector.

---

many had 5.8%, while Austria with 23.3% – behind Sweden, Latvia, and Finland – already belongs to the front runners (European Parliament, 2.4.2008).
[8] Setting on the price mechanism alone fails, as demonstrated during the doubling of the oil price within only one year.

The instruments used in the electricity sector are feed-in tariffs (FIT), premiums, quota obligation, investment grants, tax exemptions, and fiscal incentives. In the heating and cooling sector, investment grants, tax exemptions and financial incentives are used. And in the transport sector support, the instruments used are quota obligations and tax exemptions (European Commission 2011: 10).

The Commission has tried harmonisation via the introduction of the quota model but then stopped this attempt. However, the diffusion effect of support schemes can be rated. Especially the feed-in tariff is considered a most effective promotion instrument in contrast to the quota obligation. In 2011, 21out of 27 Member Countries use FIT for promoting renewably generated electricity. FIT is also regarded as a very successful instrument in other parts of the world and is therefore imported as a support scheme. International institutions, such as the feed-in cooperation initiated by Germany and Spain, could increase the diffusion effect.

The instrument of feed-in remuneration was already suggested by the first climate Enquête Commission of the German Bundestag and a few years later implemented as feed-in law through an initiative of Members of Parliament (see also Enquête Kommission 1995:1056 ff). The producers of electricity from renewable energy received a minimum remuneration and utilities were obligated to add this power into their grid. Similar promotion instruments were introduced in other European countries like for example, Denmark and Italy. However, comparison shows that the success of the instrument does not at all depend on the height of the subsidy and/or feed-in remuneration alone but instead that a number of other conditions and factors also contribute to it. These include, among other things, the licensing procedure, financing models, the tax law, public opinion, and the acceptance by residents and neighbours of wind farms or solar plants.

With the adoption of the Renewable Energy Sources Act (EEG 2000), the German electricity feed-in law was substantially expanded. It covered almost all renewable energy sources, included technology and innovation promoting incentives while maintaining the purchase obligation and guaranteed feed-in tariffs as well as investment protection for operators. Currently, 21 out of 27 EU countries apply a modified feed-in model. The quota obligation model, in contrast, belongs to the market-based instruments, whose specific aim is to promote quantity control and targets for renewable energy technology. This instrument was primarily implemented in the UK and The Netherlands, but the results were relatively minimal (for the different promotion models see Fisahn & Ptak). Nevertheless, the EU Commission has preferred this instrument for years and tried a European harmonisation of promoting renewable energies by means of the quota model. The comparison shows that success or failure of this instrument – as mentioned before for the feed-in model – depend on a number of additional factors. As with any supporting instrument, success often depends on the political frame conditions in the country. Successes of countries where effective promotion policy via subsidies and financial aid has been made over decades cannot be repeated in countries where political steering is usually done through arrangements, political agreements, and consensus mechanisms. Such national characteristics must therefore be considered.

5   IRENA – an international agency for renewables

At the same time, it seems necessary to promote renewable energies through an international institution known to enhance their potential and contribute to their global distribution.[9] The establishment of a governmental institution, however, proves to be a lengthy and contentious process. This can be reflected by the fact that renewable energy sources, even in international climate negotiations, do not have high status and that firstly only multi-stakeholder networks such as REN21 (Renewable Energy Policy Network for the 21st Century, www.ren21.net) or REEEP (Renewable Energy & Energy Efficiency Partnership, www.reeep.org) were founded. The idea of state cooperation to enhance the use of renewable energy resources is not even new. It was developed by the North-South Commission and in 1981 taken up by the Nairobi UN Conference for Renewable Energies. Several NGOs, led by Eurosolar, have been trying to bring the idea of an "International Renewable Energy Agency" (IRENA) into the negotiating process at the UN Summit on Environment and Development in 1992 (UNCED) in Rio de Janeiro.

But also ten years later, at the 2002 Summit on Sustainable Development in Johannesburg, the proposals for IRENA in the UN framework were not enforced. As before, resistance came on the one hand from oil producing countries and on the other from those oil-importing countries whose energy systems are largely based on fossil and nuclear energy. Therefore, only outside the UN an institutionalisation of renewable energy support was given a chance. A key player in this process was the German Federal Government which, in 2004, held the international conference "Renewables" in Bonn, followed by the international conferences in Beijing (BIREC 2005) and Washington, D.C. (WIREC 2008). By invitation of the Federal Government, a preparatory conference for the establishment

---

[9] In the 1950s such an institution was created for the promotion of nuclear power. The International Atomic Energy Agency (IAEA) according to its self-image worldwide promotes safe, secure and peaceful nuclear technologies (www.iaea.org).

of IRENA was held in April 2008 in Berlin. 54 countries participated and the establishment of IRENA moved a big step forward.

Finally, IRENA was officially established in Bonn on January 26, 2009 by 75 countries as an independent and international player with an efficient and slim structure and a staff of prominent experts from different geographical regions (see in detail www.irena.org). „Mandated by governments worldwide, IRENA's mission is to promote the widespread and increased adoption and sustainable use of all forms of renewable energy. IRENA's Member States pledge to advance renewables in their own national policies and programs, and to promote, both domestically and through international cooperation, the transition to a sustainable and secure energy supply".[10]

During the second session in June 2009 in Sharm El Sheikh, Egypt, the Preparatory Commission determined the Agency's interim headquarters and its Interim Director-General. The Commission designated Abu Dhabi, the capital of the United Arab Emirates, as the interim headquarters. The two other applicants still take part in the further establishment of the Agency. Bonn is hosting IRENA's centre of innovation and technology and Vienna became the Agency's liaison office for cooperation with other organisations active in the field of renewable energy.

Today, 148 states including the European Union signed the Statute of IRENA. Until April 2011 70 Signatories have ratified IRENA's treaty and became a Member of the Agency. According to IRENA renewable energy use must and will increase dramatically in the coming years, due to its key role in enhancing energy security, reducing greenhouse gas emissions and mitigating climate change, alleviating energy poverty, supporting sustainable development, and boosting economic growth. Still, the question remains whether IRENA can stimulate and which tasks it can fulfil. IRENA's main objective is to promote the use of renewable energy worldwide and to offer qualifying policy advice to member countries. Members are offered the latest scientific information, knowledge, and applied policy research. IRENA will also finance programs for the development of renewable energies and give better advice on providing the public with information.

It is thus hardly to be assumed that IRENA, being held to neutrality, will be able to bust the powerful global fossil energy networks (to their persistence forces see Altvater in this book). The structure of an international „competence centre", however, is an attempt, to strengthen and develop an international policy in favour of renewable energies beyond Kyoto, Bali, Copenhagen, and Cancun,

---

[10] http://www.irena.org/menu/index.aspx?mnu=cat&PriMenuID=13&CatID=9, accessed May, 5th 2011

where the topic expansion of renewables for the protection of the climate was only marginally negotiated. If hard „solutions" are not achievable, soft instruments are probably the only passable plan B.

6   Democracy by Decentralisation

All domains of human life depend on sufficient availability of energy sources. Renewable energies are domestic, decentralised energies. They are not only necessary for reaching demanding climate protection targets but contain the possibility of decentralised employment in cities, villages, and even smaller units with democratic potential and collective capacity for action. The employment of renewables can contribute to a development that reduces dependency from large energy companies or national suppliers, which control the use of energy sources such as the sun, wind, hydro or geothermal energy, protecting the environment and, in the end, also saving costs. The resistance on the part of the dominant energy industry, which wants to maintain ground and win in the global competition – with governmental support – particularly by capital and power concentrations and giant mergers, therefore is very strong.

Depending upon geological, climatic, and hydrologic conditions ideal conditions for a balanced energy offer, which is based on renewable energies, prevail around the globe.[11] In only one day, the sunlight reaching the earth provides enough energy to satisfy the present worldwide energy demand for eight years. Altogether, the quantity of energy from renewable sources, which could be used with the available technologies, is 5.9 times higher than the present global energy demand (Federal Government of Germany 2008). In order to open this potential, a wide promotion is necessary and apart from the choice of the most efficient locations – for instance the use of solar energy in the south and wind energy in the coastal regions – the use of all technology options. On mid-term and long-term basis, regional potential should be completely developed and fulfil targets like

- The decarbonisation, i.e. the substitution of fossil and nuclear energy by renewable energies before the last drop oil is pressed from the earth,

---

[11] The world-wide different use of renewable energies however, is today defined less by geological but rather by basic political conditions and interests: In Europe, the spectrum reaches from a 32% share of primary energy consumption in Sweden to 0.5% on Malta. In 2008, renewable energies had on average a share of 8.2% in the European Union and in Germany a share of 8.1% of primary energy consumption – a contribution for energy supply, which stands in no relation to the existing potential (REN 21, 2010).

- the penetration of strategies for sufficiency: new life styles, mobility patterns, and economy models, which are accompanied with deceleration,
- a substantial increase in efficiency by technical innovations, a more efficient energy use, reduced energy consumption, and system-adapted supply of renewable electricity (e.g. load-dependent supply, storage etc.),
- the realisation of individual and collective supply security by renewable energies (related to households),
- social justice in energy consumption,
- a transition from economic growth to qualitative development (Herman E. Daly speaks of „development without growth", 1999), and
- engaged climate protection, not only announced but implemented by challenging measures.

A full supply of the world with renewable energies will be realised faster if simultaneous and various possibilities for decentralised employment, energy conservation, and energy efficiency are gained. Because the existing energy system – at least in industrialised countries, but in developing countries as well – is based on central, fossil-nuclear technologies and yields high profits, the clarification of facts like the incontestable advantages of renewable energies and their – for human perception – infinite availability must be substantially intensified. Transformation technologies for hydro, wind, and solar energy are, in part, already well-known and represent the state of the technology. They must be offered in broad variety, become more favourable, and reach social acceptance for increased demand.

In developing countries, renewable energies offer the chance to work against the increasing energy poverty. More than 1.6 billion people do not have access to electricity. They depend on wood and manure as fuels. Here, the development of a central power supply and the appropriate net infrastructure forbids itself due to the costs alone, which would have to be carried by final consumers. In addition, the expenditure of time, which would be necessary, and the distances, which would have to be bridged and would be accompanied with high transmission losses, prohibit this way. Over 90% of generation capacities in many poorer developing countries are located in cities. On the other hand, the expansion and usage of the locally existing renewable energies could cover fast and decentralised, i.e. also in distant regions, electricity requirements of over a billion people. In addition, renewable energies have the technological advantage of flexibility in their dimensioning and they can pave the way for energy self-sufficiency in developing countries at a high speed (Scheer 2004).

In the industrialised countries, the transformation of the infrastructure is above all decisive for the energy turnaround. Ideal frame conditions for the development of renewable energies have to be created with adequate access of sustainably produced energy to existing net infrastructures, by removing any administrative obstacles or by establishing municipal self-supply. Even today, the frame conditions and accumulated subsidies – in Germany for coal and nuclear installations alone they amount to several hundred billion Euros – onesidedly favour the existing energy system. That is valid generally for all industrial and developing countries. Additional projects of the World Bank and other international banks are still aligned to the promotion of fossil energies, in particular.

But it may not be forgotten that not only private-economic actors, lobbyists, and political institutions secure the inertia of the existing energy system. Households and individuals, too, are contributing to the relative stability of the energy system. Fossilism is deeply rooted in the societies of the North and the South. Therefore, from mobility via consumption up to leisure behaviour, individual motivations and all areas of life are part of the problem. This is one reason why the employment of renewable energies cannot simply be ordered from above, but can instead only be realised through a long-term society-changing project, which has to act on all political levels: from the local up to the global.

However, there is no royal road and there may never be one. In the same way as renewable energies stand for decentralism, energy sovereignty, self-determination, and finally a way towards independence, their employment must also be implemented in bottom up-procedures. That does not exclude or forbid an international institutional stabilisation of renewable energies, national promotion programs and least of all the clarification of facts, education, and consultation. A climate-friendly and lasting global change of the energy system should much rather combine different approaches and instruments. Due to the dominant position of fossilism, which further opposes any decentralisation and transformation of the energy system, resistance and blockade attitudes must be calculated as well. Without social and political arguments for ideal solution methods „the solar world community" (Scheer 2004) is hardly realistic.

References

Amery, C. & Scheer, H. (2001): Klimawechsel. Von der fossilen zur solaren Kultur. München: Kunstmann

Brunnengräber, A. (2011): Multi-Level Climate Governance. Strategic Selectivities in International Politics, in: Knieling, Joerg, Leal Filho, Walter (2011): Climate Change Governance, HafenCity University Hamburg, Frankfurt: Springer (in prep.)

Daly, E. Herman (1999): Wirtschaft jenseits von Wachstum – die Volkswirtschaftslehre nachhaltiger Entwicklung, Pustet, Salzburg, München

DIW, DLR, ZSW, IZES (2008): Wirkungen des Erneuerbare-Energien-Gesetzes (EEG) aus gesamtwirtschaftlicher Sicht. Studie im Auftrag des Bundesministeriums für Umwelt, Naturschutz und Reaktorsicherheit, Berlin

Enquete-Kommission (1995): Mehr Zukunft für die Erde. Nachhaltige Energiepolitik für dauerhaften Klimaschutz. Schlußbericht der Enquete-Kommission "Schutz der Erdatmosphäre" des 12. Deutschen Bundestages. Bonn

European Commission (2008): 20 20 by 2020. Europe's climate change opportunity. Brussels, 23.1.2008 COM(2008) 30 final

European Commission (2009): Directive 2009/28/ EC on the promotion of the use of energy from renewable sources and amending and subsequently repealing Directives 2001/77/EC and 2003/30/EC. Brussels, 23.4.2009

European Commission (2011): Renewable Energy: Progressing towards the 2020 target. Brussels, 31.1.2011, COM(2011) 31 final

Fischedick, M. (2008): Erneuerbare Energien und Emissionshandel. In: Schüle, Ralf (Hrsg.) (2008): Grenzenlos Handeln? Emissionsmärkte in der Klima- und Energiepolitik. München: oekom, pp. 103-116

Fritsche, U. (2007): Treibhausgasemissionen und Vermeidungskosten der nuklearen, fossilen und erneuerbaren Strombereitstellung – Öko-Institut Arbeitspapier, Darmstadt

Hirschl, B. (2008): Erneuerbare Energien-Politik. Eine Multi-Level Policy-Analyse mit Fokus auf den deutschen Strommarkt. Wiesbaden: VS Verlag für Sozialwissenschaften

IEA (2008): Energy Technology Perspectives 2008 – Scenarios and Strategies to 2050, Paris

IEA (2010): 2010 Key World Energy Statistics, Paris

Kollert, R. (1994): Klimarisiken durch radioaktives Krypton 85 aus Kernspaltung, Bremen

Lindberg, L. (ed.) (1977): The Energy Syndrome; Lexington/Mass

Lönnroth, M. (1986): The Coming Reformation of the Electric Utility Industry, in: Johansson; et al (Eds.), Electricity, Lund 1986, pp. 765-786

Meadows, D.; Meadows, D. L.; Randers, J.; Behrens, W. W. (1972): The Limits to growth; a report for the Club of Rome's project on the predicament of mankind, New York: Universe Books

Meadows, D. H.; Meadows, D. L./Randers, J. (1993): Beyond the Limits: Confronting Global Collapse, Envisioning a Sustainable Future, Chelsea Green Publishing

Mez, L.; Osnowski, R. (1996): RWE – Ein Riese mit Ausstrahlung, Köln: Kiepenheuer & Witsch

Mez, L.; Piening, A. (2006): Phasing-Out Nuclear Power Generation in Germany: Policies, Actors, Issues and Non-Issues, in: Jänicke, M.; Jacob, K. (Eds.), Environmental Governance in Global Perspective. New Approaches to Ecological Modernisation, Berlin: Freie Universität Berlin, pp. 322-349

Mez, L.; Schneider, M.; Thomas, S. (eds.) (2009): International Perspectives on Energy Policy and the Role of Nuclear Power, Brentwood: Multi-Science Publishing

REN21 (2010): Renewables 2010 Global Status Report, http://www.ren21.net/Portals/97/documents/GSR/REN21_GSR_2010_full_revised%20Sept2010.pdf, accessed April, 7th 2011

Rosenkranz, G. (2006): Mythos Atomkraft. Über die Risiken und Aussichten der Atomenergie. In: Heinrich Böll Stiftung (Hrsg.): Mythos Atomkraft: Ein Wegweiser. Berlin: Heinrich Böll Stiftung, S. 11-52

Scheer, H. (2004): The Solar Economy: Renewable Energy for a Sustainable Global Future, London: Earthscan

Schneider, M. et al. (2007): Residual Risk – An Account of Events in Nuclear Power Plants Since the Chernobyl Accident in 1986, Brussels, May 2007. Available at http://www.greens-efa.org/cms/topics/dokbin/181/181995.pdf, accessed April, 7th 2011

Federal Government of Germany (2008): The Case for an International Renewable Energy Agency (IRENA), Preparatory Conference for the Foundation of IRENA Berlin. Available at http://www.wcre.de/en/images/stories/The_case_for_IRENA.pdf, accessed April, 7th 2011

Traube, K. (2005): Atomenergie – unverantwortliche Bedrohung, marginale Potenziale, in: SPD, Atomausstieg – innovativ, nachhaltig, sicher, sozial, zukunftsweisend, Berlin, pp. 12-22

Turner, G. (2008): A Comparison of The Limits to Growth with Thirty Years of Reality. Available at http://www.csiro.au/files/files/plje.pdf, accessed April, 7th 2011

Neu im Programm Politikwissenschaft

Elemente der Politik

Hrsg. von Bernhard Frevel / Klaus Schubert / Suzanne S. Schüttemeyer / Hans-Georg Ehrhart

Blum, Sonja / Schubert, Klaus
Politikfeldanalyse
2., akt. Aufl. 2011. 198 S. Br. EUR 16,95
ISBN 978-3-531-17276-7

Dehling, Jochen / Schubert, Klaus
Ökonomische Theorien der Politik
2011. 178 S. Br. EUR 16,95
ISBN 978-3-531-17113-5

Dobner, Petra
Neue Soziale Frage und Sozialpolitik
2007. 158 S. Br. EUR 12,90
ISBN 978-3-531-15241-7

Frantz, Christiane / Martens, Kerstin
Nichtregierungsorganisationen (NGOs)
2006. 159 S. Br. EUR 14,90
ISBN 978-3-531-15191-5

Frevel, Bernhard
Demokratie
Entwicklung – Gestaltung –
Problematisierung
2., überarb. Aufl. 2009. 177 S. Br. EUR 12,90
ISBN 978-3-531-16402-1

Fuchs, Max
Kulturpolitik
2007. 133 S. Br. EUR 14,90
ISBN 978-3-531-15448-0

Jahn, Detlef
Vergleichende Politikwissenschaft
2011. 124 S. Br. EUR 12,95
ISBN 978-3-531-15209-7

Jaschke, Hans-Gerd
Politischer Extremismus
2006. 147 S. Br. EUR 14,95
ISBN 978-3-531-14747-5

Johannsen, Margret
Der Nahost-Konflikt
2., akt. Aufl. 2009. 167 S. Br. EUR 16,95
ISBN 978-3-531-16690-2

Kevenhörster, Paul / Boom, Dirk van den
Entwicklungspolitik
2009. 112 S. Br. EUR 12,90
ISBN 978-3-531-15239-4

Kost, Andreas
Direkte Demokratie
2008. 116 S. Br. EUR 12,90
ISBN 978-3-531-15190-8

Meyer, Thomas
Sozialismus
2008. 153 S. Br. EUR 12,90
ISBN 978-3-531-15445-9

Schmitz, Sven-Uwe
Konservativismus
2009. 170 S. Br. EUR 16,90
ISBN 978-3-531-15303-2

Erhältlich im Buchhandel oder beim Verlag.
Änderungen vorbehalten. Stand: Juli 2011.

www.vs-verlag.de

VS VERLAG

Abraham-Lincoln-Straße 46
65189 Wiesbaden
tel +49 (0)6221.345 - 4301
fax +49 (0)6221.345 - 4229